Lecture Notes in Social Networks

Series editors

Reda Alhajj, University of Calgary, Calgary, AB, Canada
Uwe Glässer, Simon Fraser University, Burnaby, BC, Canada

Advisory Board

Charu C. Aggarwal, IBM, T.J. Watson Research Center, Hawthorne, NY, USA
Patricia L. Brantingham, Simon Fraser University, Burnaby, BC, Canada
Thilo Gross, University of Bristol, Bristol, UK
Jiawei Han, University of Illinois at Urbana-Champaign, IL, USA
Huan Liu, Arizona State University, Tempe, AZ, USA
Raul Manasevich, Universidad de Chile, Santiago, Chile
Manasevich Anthony J. Masys, Centre for Security Science, Ottawa, ON, Canada
Carlo Morselli, University of Montreal, QC, Canada
Rafael Wittek, University of Groningen, The Netherlands
Daniel Zeng, The University of Arizona, Tucson, AZ, USA

More information about this series at http://www.springer.com/series/8768

Sorin Adam Matei • Brian C. Britt

Structural Differentiation in Social Media

Adhocracy, Entropy, and the "1 % Effect"

 Springer

Sorin Adam Matei
Purdue University
West Lafayette, IN, USA

Brian C. Britt
South Dakota State University
Brookings, SD, USA

ISSN 2190-5428   ISSN 2190-5436   (electronic)
Lecture Notes in Social Networks
ISBN 978-3-319-87791-4   ISBN 978-3-319-64425-7   (eBook)
DOI 10.1007/978-3-319-64425-7

Printed on acid-free paper

This Springer imprint is published by Springer Nature
The registered company is Springer International Publishing AG
The registered company address is: Gewerbestrasse 11, 6330 Cham, Switzerland

Acknowledgments

This book is the product of a long journey. Although there are only two authors on the cover, the book and the projects behind it are the product of a true fellowship. We are in debt to the many people and many organizations that made this book possible. First, we would like to thank Purdue University and the National Science Foundation, specifically those officers, administrators, and anonymous reviewers and colleagues who believed in our project and in our work. Several Purdue grants, including those awarded by the College of Liberal Arts, by the Purdue Research Foundation, by the Office of the Vice President for Research, and by Discovery Park, made this book possible, and the NSF award BCS 1244708 was instrumental in bringing this project to fruition.

Among the many individual people who selflessly supported this book, we need to first acknowledge Elisa Bertino, the head of the Purdue University Cyber Center and a mentor for us all in grant writing and research design. Luca de Alfaro, from the University of California, Santa Cruz, graciously shared the core data set of Wikipedia revisions for the 2001–2010 interval with us, including a core metric for effort at the revision level. We have also benefitted from support from our colleagues, Michael Zhu and Chuanhai Liu, as well as from their graduate student, Wutao Wei. Another of our colleagues, David Braun, and our graduate students, David Lazer and Azat Khairov, also contributed a good amount of effort through various channels, including the Discovery Park Internship Programs. Serendipity Labs, in Chicago, was the place where this volume took shape, and the hospitality of this co-working space and of its manager, Andreas Brandl, should also be mentioned.

We are also in debt to our families and friends who listened to our presentations and encouraged us along the way. These individuals are too numerous and their support too great to properly detail here, but this volume, and the years of work that culminated in it, would not have been possible without their love and the strength that they gave us.

Contents

Part III   Future Theoretical and Practical Directions

About the Authors

Sorin Adam Matei is a Professor, Brian Lamb School of Communication, Purdue University. Dr. Matei studies the emergence of new forms of sociability online. His papers and books examined the emergence of social capital online, the effect of offline sociability online on online relationships, and the generation of trust and authority in knowledge markets. His work was funded by the National Science Foundation, Kettering Foundation, and Mellon Foundation.

Brian C. Britt Assistant Professor, Department of Journalism and Mass Communication, South Dakota State University. Dr. Britt is a computational social scientist who focuses on the intersection between organizational communication and new media, with a particular emphasis on the strategies employed by individuals positioning themselves in online organizations. His work has been funded by the National Science Foundation and the National Institutes of Health and has resulted in several book chapters in edited volumes.

Chapter 1
Introduction

1.1 Inequality in Online Groups

If Wikipedia was a country and words its income, the wildly uneven distribution of wealth among its contributor-citizens would be unheard of. The wealth distribution of the USA, for instance, which some see as quite unfair, pales by comparison, as the top 1% of American earners account for "only" 15% of the national income compared to the 77% of content "owned" by the top 1% of Wikipedians (see Fig. 1.1).

Our book explains why and how this uneven distribution matters. We believe that, far from being an abnormality, this inequality is a sign that Wikipedia has developed organically and naturally. The imbalance, in fact, shows that the site is structurally differentiated and that functional leaders have emerged, both processes that are essential for a healthy, functioning project. In this respect, the proverbial top 1% of contributors have a positive effect, as they anchor and serve as a contributing elite group with an important role in shaping the project.

We call the emerging Wikipedia elite "functional leaders." Unlike elected or nominated leaders, or leaders who are hired and bestowed with certain rights and privileges consecrated by an organizational chart and bylaws, functional leaders lead by doing. They do more and contribute more frequently than anyone else in the same project. They are the most productive contributors. They naturally emerge in many social media spaces, initially as essential nodes of production and later through leading contributions as well as their interactions with other members.

In this volume, we analyze the emergence of functional leaders on Wikipedia to explain how this elite group shapes one of the most successful social media projects on the planet in terms of both content and organizing the division of labor. In the process, we also show how the Wikipedia elites tend to become stable and increasingly influential over time. Finally, we interpret these findings, suggesting that social media projects that have a collaborative goal develop, in a rather short period, a new, resilient, yet different type of social hierarchy that combines features of

© Springer International Publishing AG 2017
S.A. Matei, B.C. Britt, *Structural Differentiation in Social Media*, Lecture Notes in Social Networks, DOI 10.1007/978-3-319-64425-7_1

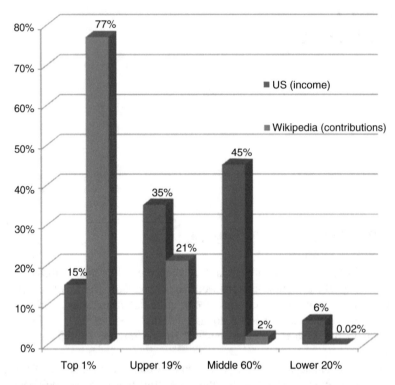

Fig. 1.1 If Wikipedia was a country, how equitable would its wealth distribution be? The top 1% Wikipedia contributors dwarf the top 1% income earners in the USA. The former group contributes 77% of Wikipedia's content, a figure five times as large as the latter's share of the national income

traditional and new, online, social organizations. The main proposition is that these features do not resemble anything that we have previously observed. They are not authoritarian and top-down, nor democratic, nor egalitarian. They are a form of adhocratic, hierarchical social organization. To explain the terms and its evolution, we propose a new organizational change framework, which considers a gamut of organizational configurations and transition processes that drive their evolution.

As a brief introduction, the term "adhocracy" suggests that, far from being decentralized, spontaneously coordinated crowdsourced projects such as Wikipedia are, in fact, orchestrated by an organizational system that combines a stable power hierarchy with individual mobility. More experienced members maintain a relatively structured collaboration process, while—at least on Wikipedia—users ascend to positions of power through an osmotic process of social advancement. Our time series analysis of Wikipedia contributions shows that as the elite group becomes more stable, with the same individuals maintaining their elite status over time, the project tended to reach and stay at a relatively high level of social structuration. In addition, we used a social entropy method to measure social structuration by observing the inequality of contributions. This approach showed that over time, Wikipedia editors tend to separate into subgroups representing varying contribution intensity levels and functional roles, and these subgroups maintain considerable stability over time.

One especially distinguishing feature of our book is its broad set of theoretical and practical implications. In this book, we explain the prevalence of uneven contribution distributions across online projects from a new perspective. We reject the claims that these skewed distributions merely result from the fact that not all contributors can work together all of the time. Rather, we demonstrate that top contributors tend to form a stable elite through which they build and maintain their roles. In addition, we propose an organizational configuration model that advances the concept and operationalization of structuration from a unidimensional measure to a five-dimensional space describing organizational change.

1.2 Theoretical Starting Points

The advent of the Internet promised faster and more direct social interactions. Some predicted more equal contributions and greater access to the human store of knowledge (Benkler 2006; Brafman and Beckstrom 2006). Yet, empirical observations have dampened these claims of egalitarianism, suggesting an alternative scenario: inequality is just as prevalent in online interactions as it is in real life, if not even more so (Matei et al. 2015).

Shirky (2008) offers an especially lucid assessment of this inevitable imbalance. He proposes that the amount of voluntary work invested in Internet communities follows the Pareto principle (also known as the 80/20 rule) wherein a few individuals (20%) contribute most (80%) of the effort. While the exact proportions might vary, particularly in social systems with even more extreme long-tail distributions (Huberman 2003), the observed inequality is itself not unusual. Barabási (2014; see also Newman et al. 2006), among others, referred to it when discussing the power-law distribution of web links and interactions. It has also been observed in many other human pursuits, particularly the distributions of income, of volunteer time, and of work team effort, the latter of which will be discussed below.

With a few exceptions, such as Shirky's work and more empirically based research (Kuk 2006; Ortega et al. 2008) on free and open source software (FLOSS), as well as the occasional paper dedicated to Wikipedia (Muchnik et al. 2013), the uneven distribution of inputs and outputs in social media remains a topic that leaves many questions to be asked and answered. Of course, we know and expect that some voices might dominate social media discourse. Yet, we are only beginning to understand some of the most basic differences between the elites who dominate social media and those at the top of the income, civic, and other social distribution pyramids. In particular, despite the revelation that online volunteer contribution distributions tend to be much more skewed than those of face-to-face social systems[1]

---

[1] "[A] tight-knit community of 600 to 1000 volunteers does the bulk of the work [on Wikipedia], according to Wikipedia cofounder Jimmy Wales."

(Shaw 2008), we lack an integrated theory-based perspective to explain *why* the online realm in particular lends itself to such dramatically uneven distributions.

Beyond that, examining the extreme inequalities of online groups begs a few additional questions about the nature of elites in such communities, especially since they are set apart from the masses to a much greater degree than the top participants in most face-to-face groups. For instance, how does one become a super contributor to a platform like Wikipedia and how long does one's status as such last? More importantly, how does the presence of such contributors influence the system-level contribution dynamics? Finally, are organizations with such significant contribution inequalities stable? If not, how do they evolve? The evolutionary perspective is particularly important for our study, as it grounds the discussion about social media projects on a principled sociological basis, connecting older concerns with newer theories about social organization among humans.

Beyond the broader theoretical considerations, the current study is grounded in a very specific domain of investigation, addressing the nature and extent to which contributions made to the online encyclopedia Wikipedia are unequal. Yet, describing the extent of inequality is not the ultimate goal. The final goals are:

1. To longitudinally describe elite contributions and the process of social differentiation
2. To derive an empirical model for evolutionary differentiation
3. To use elite evolution, both in terms of resilience and relative weight in the production system, as an explanatory lens for the global evolution of an online (social media) system
4. To demonstrate that social media dynamics lead to a new type of leadership borne out of an adhocratic organizational structure that combines top-down and bottom-up processes
5. To propose a principle-based evolutionary model for organizational change
6. To develop a multidimensional method to explain transitions between organizational configurations

Our theoretical framework is that of social differentiation enhanced with an adhocratic perspective (Matei and Bruno 2014). Through it, we argue that the unevenness observed in many social media spaces conceals an even more important process: the emergence of division of labor and a hierarchical functional power structure that is, at the same time, adhocratic. By this, we mean that while the identities of the roles and missions performed by those at the top of the social structure are quasi-stable, the occupants of those roles and performers of those missions are subject to highly contingent factors such as availability, social interaction constraints arising from time and space, and so forth.

Building on this argument, we propose an evolutionary model complete with a new theory for the emergence of a small and resilient contributor elite on Wikipedia. Highly active contributors effectively shape the social structuration of the project. They create and perform specific functional roles, especially that of the "elite contributor," an individual who leads by example. He or she shapes the collaboration and the content by intervening often, at certain intervals, to shape the overall content of an article or topic.

We also propose a method to detect how long elite members survive at the top of the contribution pyramid, which is subsequently used to explain how resilience or "stickiness" among members of the elite group influences the emergence of functional leadership. Beyond this, we evaluate the ways in which individual connections to members of the elite create a path for advancement to an elite contributor status. In other words, we consider whether an ordinary contributor's collaboration with elite members affects his or her likelihood of advancing from ordinary to elite status as well.

Finally, we propose a synthesized theoretical model of organizational change that integrates five dimensions of collaborative activities on Wikipedia. The model incorporates the development of collaborative attractiveness, extroversion, communication flows, structural order, and trends in the choice of coeditorial partners. Our goal in using this model is to determine the specific historical points at which Wikipedia changed organizational configurations between set of several possibilities: entrepreneurial, machine bureaucratic, professional bureaucratic, divisionalized, and adhocratic. Our explanatory model can naturally be extended to other social media, explaining how they develop a specific type of social order within which social mobility, power differentiation, and division of labor are the norm.

In the end, we aim to contribute to the broader understanding of the purported "egalitarianism," or lack thereof, in social media projects and how much a new class of "sticky elites" shapes the fate of social media projects.

1.3   Problem Significance

It is generally easy and cheap to join and to add content to social media sites such as Wikipedia. After all, this is part of the new generation of so-called user-generated media. Wikipedia is built around the idea that anyone who wants to change content should be able to do so. Because of such functionality and inclusiveness, many believe that social media like Wikipedia are spaces of free association and minimal hierarchically driven interaction (Gillmor 2006; Goode 2009).

Social media are indeed peer production hubs in which individuals connect to each other without supervision or censorship, and users are clearly not controlled or directed by a centralized formal control mechanism. And, to a certain extent, social media users do co-orient their contributions based on the cues that are provided by the tools they use, engaging with the content in a somewhat similar manner, thus suggesting a form of homogeneity across users.

With that said, is this co-orientation the only coordination mechanism existing in social media spaces? Are social methods of coordination also important? To what degree do these ultimately encourage more or less evenly distributed contributions? Further, how much do these methods clash with technological coordination mechanisms, which some believe discourage the emergence of formal social roles and leadership positions (Benkler 2006; Segerberg and Bennett 2011; Shirky 2008)?

We suggest that Wikipedia and other social media do involve ad hoc leadership structures that are responsible for most social coordination. These are not immediately apparent or even immediately present in social media projects. They evolve over time and are, in certain instances, preceded by organizational configurations that resemble older and more common forms of organizing. Yet, in the end, social media tend to foster a certain kind of leadership that is defined by the amount of work completed rather than by an official organizational chart. Such leadership plays a functional social role, that of funneling effort and content in particular ways, which may exacerbate unequal social contribution patterns.

To preface the more detailed argument to be presented below, let us briefly examine how the skewness of contribution patterns on a social medium like Wikipedia compares to those found in more mundane areas of human activity. For the sake of brevity, we will only mention civic and economic systems as illustrative examples, but many others could be invoked as well.

We analyzed the entire Wikipedia editorial database, containing every single edit made to every article, between 2001 and 2010, for a total of approximately 250 million edits made by roughly 22 million editors. Three million of these editors were attributed to logged-in accounts, while the remaining 19 million were anonymous users who were nonetheless identifiable by IP addresses (see below for methodological details). Our study starts from one foundational finding of our own analysis: the top 1% of all Wikipedia contributors are responsible for 77% of the collaborative effort on Wikipedia based upon the extent to which the text of articles was actually changed. Likewise, the top 20% of the editing population covers a staggering 99% of Wikipedia's collaborative effort.

This far exceeds the comparatively modest inequalities among civic activities in the USA. For example, the American Time Use Survey conducted by the Bureau of Labor Statistics (2016) reveals that between 2003 and 2011, 1% of the US adult population was responsible for 40% of the total time volunteered to civic, religious, or sport associations.[2] This is an impressive enough figure, but it is nonetheless only half as overwhelming as the contribution inequality observed on Wikipedia. Further, if we only take into account the people who volunteered at least once (much like how our Wikipedia figures only consider individuals who revised an article at least once), the top 1% of all volunteers contributed only 5% of all volunteer time, and individuals in the top 20% were responsible for merely 50% of the time volunteered in the USA.

Inequality on Wikipedia is also far more intense than that of income distribution. The top 20% of the American workforce, each of whom makes over $101,582 per year, jointly reap 51% of the combined national personal income (US Census Bureau 2014). Again, although this figure may be noteworthy in its own right, it pales in comparison to the 1% of Wikipedia contributors responsible for 77% of the work.

---

[2] Our calculations were derived from the Bureau of Labor Statistics' primary data.

The sharp divergence between inequality online and the distribution of traditional social interactions demands an explanatory mechanism. Shirky's (2008, 2010) popular books on the social media revolution opened the conversation. Shortly thereafter, Kumar et al. (2006) highlighted the prevalence of this inequality in online social networks, and Ortega et al. (2008) demonstrated its applicability to free software projects (FLOSS). In a limited sense, using simpler measures than the ones proposed here, Ortega et al. (2008) showed that the online community of Wikipedia editors does indeed have an unequal distribution of contributions. The more important question, however, is not merely whether or not such inequalities exist, but rather, what they mean.

Some early work (Javanmardi et al. 2009; Nov 2007) attributed online inequalities to the different motivations and abilities of contributors. Matei and Bruno (2014) and Arazy and Nov (2010), however, extended the conversation about the nature of such differentials, arguing that inequalities on Wikipedia represent much more than natural variations of individual effort or abilities. Matei and Bruno (2014) in particular suggested that inequalities tap into something more significant: the emergence of leading functional roles. Such role differentiation speaks about the fundamental nature of human collaboration.

At the same time, the ethical dimension of unequal distributions remains important, as examining inequality in social media may yield a better understanding of equality across contexts. Equality, after all, is the cornerstone of democratic society. Are online processes a reflection of our existing (presumably) egalitarian society? Or are they making an already unbalanced situation even worse?

1.4   Research Strategy

To answer such questions, we need to move beyond the descriptive research that currently predominates. The next natural step is to propose explanatory models for how online inequality emerges and why it matters. To this end, we have developed a social structural evolutionary model in which observed inequalities in social media are considered to be symptoms of larger social growth and transformation. In short, when we detect inequalities, it means that loosely connected individuals are about to form more complex communities of practice. Such inequalities also suggest the emergence of functional roles and of a stewardship mechanism to acculturate newcomers. When observed in a multidimensional space that incorporates a variety of interactional choices, we can describe inequality as a part of a set of interactional dynamics that correspond to specific organizational configurations.

In this volume, we propose and follow a multistep research strategy for deploying our explanatory mechanism:

1. Map the evolution of inequality on Wikipedia and detect the distinct growth phases determined by the emergence of functional leaders over a period of 9.5 years (2001–2010).

2. Develop and test an organizational evolution model for Wikipedia.
3. Study the in-project social mobility of functional leaders on Wikipedia by measuring the stability of the top 1% of all Wikipedia contributors over time as long-term members of this elite group.
4. Hypothesize a hierarchical model of functional elites that accounts for both top-down and bottom-up processes.
5. Propose an online organizational transformation framework that relies on a multidimensional organizational configuration approach.
6. Validate the proposition that Wikipedia advanced toward an adhocratic organizational model over time.

Of these three steps, 4 and 6 are perhaps the most important, as they address processes that have largely been ignored in the literature: namely, the role of social influence in the formation of digital elites and paths for organizational transformation. At the same time, our strategy also promotes a new way of understanding the social evolution of the online spontaneous groups that are responsible for much of the knowledge that we use in our everyday lives.

1.5   Theoretical Perspective

1.5.1   Inequality and Evolutionary Processes

Besides structural differentiation, another cornerstone of our research is an evolutionary perspective on organizational growth within online projects in which members may freely associate with one another. A number of theoretical perspectives support the proposition that in spontaneously emerging organizations, contribution inequality is not an isolated phenomenon but is instead fundamentally connected to other organizational dynamics. We employ a distinct theoretical strand that treats this inequality as a symptom of, among other things, functional differentiation (Bales 1950; Blau 1977; Britt 2013). While inequalities could certainly be exacerbated by authoritarian control and manipulation, a significant degree of inequality emerges during the evolution of online collaborative communities via their natural process of social organization alone. This is also what we would expect to find on Wikipedia, which experiences relatively minimal formal managerial intervention.

In short, as spontaneously emerging organizations take on more complex tasks and as their members develop ongoing interactions over time, roles and a division of labor emerge. This naturally creates an unequal distribution of inputs and outputs, which thus reflects a process of structural differentiation. Structural differentiation, in turn, develops over time as the organization proceeds through a series of different configurations.

An evolutionary perspective represents the optimal avenue for explaining inequality as functional and structural differentiation. After all, inequality can be seen as a component of functional differentiation if, when mapping it over time, we uncover discrete phases and emergent social roles (see Britt 2013) and, at the same time, identify specific influences that drive transformations into various organizational configurations. This perspective is congruent with the evolutionary paradigm proposed by Wenger (1998) for communities of practice, some of the most conspicuous forms of spontaneously emerging organizations. The community of practice lens also fits Wikipedia, our object of study, as suggested by Ayers et al. (2008) and O'Sullivan (2009). Our organizational configuration change model, likewise, builds upon the foundations of several long-standing theoretical domains, including the seminal works of Van de Ven and Poole (1995) and Mintzberg (1979).

Communities of practice are domain-specific voluntary associations with a practical orientation. They generate, exchange, and curate "practice," that is, know-how informed by knowledge. Structurally, they are crosscutting communities based on functional roles. In such communities, individual participants focus on the most sensible ways of generating a particular outcome (product, process, know-how, knowledge, etc.) and perform their functions accordingly. It is not especially important for members of such groups to follow formally prescribed modes of production or official chains of command, yet leaders still serve a vital role in guiding the work effort. These leaders are rarely appointed; rather, they naturally emerge from the process of collaboration itself, leading by doing and stewarding by showing newcomers the standard practices and norms of the community—in other words, demonstrating "how it's done."

Ayers et al. (2008) provide a historical perspective to illustrate how Wikipedia functions as a successful community of practice. Briefly, the community relies on informality and bold leadership. It fosters functional, achieved roles, eschewing formal hierarchies and ascribed organizational roles. Yet leaders are nonetheless real and influential. They represent the future of the project, teaching new members how to use the tools and showcasing the norms related to editing, organizing content, and combating vandals. While many editors move between these roles, there are others who specialize in one particular task. In short, much like the old Royal Society of London, Wikipedia is a community dedicated to the practice of gathering and sharing knowledge. Their organizational mechanisms may differ, but their goals are the same.

According to Wenger (1998), communities of practice typically start with a rough period of incubation. During this phase, members experiment with the tools and rules and try to define their own participant roles. This is a time of sudden shifts in responsibilities and membership. Contributions can be fairly even at times, while at other points certain members dominate.

The second period is that of coalescence, a stage in which the community emerges as a well-defined collaborative space. Members negotiate the manner in which the community will operate, developing initial behavioral norms. This slowly reduces the initial chaos of the growing social system. During this period, the ultimate direction in which the organization will grow is clarified, at least to a degree. The long-term community goals and an envisioned trajectory for reaching them also become more tangible.

In the third, climactic phase, associative processes reach a form that the organization would consider close to its ideal. This is also when a final stewardship system emerges. Specific members take charge of both the coordination and the contribution process for longer periods of time. Inequality also stabilizes once the leaders have emerged. Structural differentiation, whereby functional leaders become distinguishable from ordinary participants, is complete. Activity may expand or decay over time, depending on the effectiveness of the leaders.

Britt's (2013) initial organizational-level examination of Wikipedia falls in line with Wenger's claim. According to Britt, the first phase of Wikipedia's development was a period of gestation during which inequality varied wildly and even declined to a certain degree. He equated this period with that of a simple organization, during which Wikipedia relied on the efforts of a very small number of contributors whose participation was sometimes inconsistent. Although Britt primarily focused on high-level organizational structures rather than the fine-grained behaviors of individual editors, at this stage we can nonetheless conjecture that the community appeared to be exploring various schemes for self-organization and leadership roles. In any case, after reaching a relative minimum, egalitarianism on Wikipedia rebounded until eventually stabilizing, at which point it remained at a rather high level for years afterward. This signaled that the organization had finally settled on a more permanent structure and suggested that organizational leaders and stewards had also emerged. The structure was, however, an adhocratic one, as we will show in the text to come.

One of the purposes of this manuscript is to expand on this initial model and to provide a comprehensive method to define and track organizational change. This approach is detailed at length in part II of this volume.

In brief, the evolutionary model for communities of practice proposes specific phases of inequality, with discrete inflection points separating each phase from the other. This volume illustrates how to detect such points in the evolution of inequality on Wikipedia. In addition, it proposes an organizational configuration change model, which overlaps and more precisely describes the evolutionary process.

By detecting the phases in the evolution of inequality, we also facilitate a further exploration of the emergence of functional leaders. As previously noted, a steady state of inequality indicates that a long-term stewardship mechanism has emerged, through which certain individuals willfully lead the coordination efforts. These stewards invest greater amounts of energy and time in the project and set the tone and expectations for the larger working group, becoming functional leaders. They lead by example.

Wenger (1998) and others (see, e.g., Bales 1953) argue that the emergence of stewards is a natural and recurrent process. The roots of this perspective can be traced to the earliest studies in organizational sociology. Durkheim's (1893) *Division of Labor in Society* is the earliest investigation of the issue, and several of its core concepts, especially roles and dynamic density, are still useful today. With that said, Blau's (1977) more recent foundational work is also quite relevant. He argued that we need to consider inequality (in terms of organizational work) as a binding factor when a given organization exhibits sufficient heterogeneity in its group affiliations, a supposition that is quite germane in this context.

Bales' (1953) work on the distribution of effort in small teams can also be invoked here. Bales observed that in most spontaneous small teams, a few members are ultimately responsible for the majority of a team's output. In his classic study on small group interaction, Bales noted that, across experiments, the distribution of effort was unequal. Even in groups of complete strangers with no prior history, some members stood out. They interacted more, talked more, and were recognized as leaders more often than other members. This illustrates, firstly, the prevalence of role divisions and contribution inequality across contexts. For our purposes, it also further explains why such obviously critical inequalities might be even more pronounced in some contexts—like Wikipedia—than others.

1.5.2 Adhocracy: Social Mobility Moderates Inequality

The suggestion that uneven contributions are symptoms of structural differentiation demands a caveat, as inequality might have two main components. The first of these would be social/interaction inequality, a mere measure of skewness in the distribution of rewards and benefits across the system at various points in time. The second is social mobility, a measure of how "sticky" people are in receiving this skewed proportion of rewards and benefits over time—in other words, whether elites and followers tend to resiliently maintain their relative statures from day to day, week to week, and so forth.

This distinction is crucial. Even if inequality in a system is high at a particular point in time, the community can be characterized by even contributions *over time* if mobility is also high. When everyone rotates in and out of the "top" positions over time, inequality matters less. In fact, in such a scenario, structural differentiation may be low or almost nonexistent. Furthermore, inequality at a particular moment may be a mere statistical artifact if contributors are simply taking turns in adding content to a given project. Although some people do contribute a large amount of content within a given temporal span, such as a day, a week, or a month, the supposedly unequal contributions might even out as different people contribute at different times. Yet, if specific community members consistently contribute the most over a long duration, then those unequal distributions become true footprints of functional roles played by certain "sticky" contribution leaders. All the same, even if functional leaders emerge, it is not wise to presume that top contributors rigidly or oppressively dominate the contribution process. Organizations, especially those that function online, experience considerable turnover. As voluntary groups with low barriers to entry, online group membership will inevitably vary over time.

The research literature reviewed by Britt (2013) on the pattern of contributions to social media suggests an alternative, middle-ground view. While inequality is not a mere statistical artifact (Matei et al. 2015), there are nonetheless indications that the resilience of the functional elites varies over time. On this basis, we can expect a mixture of both elite resilience and elite membership turnover. Perhaps the best summation of this phenomenon comes from research on organizational leadership dynamics and, in particular, the idea of an adhocracy, an organizational configuration that serves as an especially apt description of Wikipedian processes. After all, the name adds a nuance of fleetingness to an otherwise structured, long-lasting process of leadership.

Let us address this concept at greater length. The term "adhocracy" was first introduced to the social scientific vocabulary by Alvin Toffler (1970) in his work on postmodern organizations. Mintzberg (1979, 1989) and Waterman (1993) later adapted the term to organizational science, using it to describe leadership in the context of just-in-time production and customized consumption. More recently and less formally, "adhocracy" was defined in *A Dictionary of the Internet* (Oxford Reference 2009) as a form of organization "that [does] not rely on job descriptions, hierarchy, standards, and procedures; rather, workers in the company carry out tasks because they need to be done."

Adhocracies come with an important substantive proposition that goes beyond semantic innovation. Positions of power exist only as "virtual loci" of control. They are not substantive, permanent functions to be occupied by specific individuals for an indefinite period of time. Instead, they are relatively easily seized and, after a period, which might be prolonged but not unbounded, relinquished. In view of this, it is important to identify the rate and the mechanisms by which positions are gained and surrendered. More importantly, we need to determine how the top members or contributors survive in the elite group and how long they do so. In other words, what is the individual and group stickiness of the elite and by what methods do members enter or leave the elite group?

One core mechanism by which adhocracies promote leaders is osmotic inclusion. New members are slowly and individually acculturated into the environment. The process is not one of deliberate recruitment but a gradual progression from user to leader, as suggested by Preece and Shneiderman (2009). Future elite members are only dimly aware of their initial own position in the grand scheme of things. They engage in tasks and incrementally increase their effort, and they become elite members as their productivity reaches a loosely defined threshold. They take over for the members who tacitly drop out as the priorities, tasks, and time availabilities change. The process is one of overlapping individuals, microgroups, or dyads that naturally and seamlessly exchange roles and in-progress work activities over time, and it has no formal boundaries or mechanisms. In short, users ascend to the elite merely by working harder than everyone else.

Yet, the likelihood of stepping over the threshold may be enhanced through active engagement with existing elite members. Theoretically, a relatively productive nonelite member might have a higher chance to ascend to an elite position if he or she more closely collaborates with an elite member, and likewise, the pace of advancement to leadership positions could be hastened by establishing connections with existing elites, thereby growing more acculturated more quickly.

In our study, one core analytic step is detecting and mapping the longitudinal changes in the likelihood of nonelite users to become members of the functional elite on Wikipedia. In the very specific terms of our study, we test the hypothesis that one's likelihood of joining the group of functional leaders on Wikipedia is indeed dependent, at least in part, on the establishment of prior collaborative connections with existing leaders. Our findings further open a secondary discussion about the adhocratic nature of the leadership cadre.

One note of caution should be issued about the interpretation of the adhocratic mechanism as described above. While adhocracy can explain, to a certain extent, the emergence of functional leadership positions, it cannot explain everything. Wikipedia is far more complex than the terms "functional role" or "osmotic induction" alone might suggest. In particular, the site includes several layers of formal control mechanisms, enforced by formal roles, including the "bureaucrats," "admins," and "sysadmins." The former two groups have administrative rights, while the latter has coding privileges and access that allow them to run the software and hardware that power Wikipedia. These privileges include the ability to add, eliminate, and change site functionalities and utilities, rights that are exclusive to the formally defined role and which shape the entire contribution process. No study of Wikipedia's collaborative structure can be considered complete in the absence of an acknowledgment of formal leadership roles. In that respect, the present study is only a partial attempt to disentangle one power and social structuration mechanism among many on Wikipedia. Although functional leadership is critically important on Wikipedia, that is not the only type of leadership at hand.

1.5.3 Wikipedia as a Site of Investigation: Significance, Research Review, and Data

As a whole, this volume examines Wikipedia as a major social project. Regardless of any arguments about the quality (or lack thereof) in its content (Giles 2005; Mullan 2009), Wikipedia's prominence and relevance to this text come from its central role in the everyday lives of the general population. Briefly, Wikipedia is the fifth most visited website in the world (Alexa 2016). Approximately 300 million individuals check Wikipedia each day, a figure that equates to just under 10% of the world's Internet-connected population of 3–3.5 billion people (International Telecommunication Union 2016). Just as importantly, Wikipedia is joined at the hip with major social media and search technologies. At least eight in ten searches for common nouns on Google return a Wikipedia article as one of the top 10 results (Miller 2012; Silverwood-Cope 2012). Consequently, Wikipedia is one of the most common sources of information for student papers, with over 80% of college students making use of it (Head and Eisenberg 2010; Colón-Aguirre and Fleming-May 2012), and it is even becoming a major source for journalists (Cision and Bates 2009), as 61% of professionals use it for their own research (Shaw 2008).

Wikipedia is, however, more than a simple reference site: it is a community of contributors in continuous growth. It grew from a dozen hackers and enthusiasts, bandied together by a former commodities trader and online entrepreneur, into a community of over 20 million registered users. We should add that tens of millions of anonymous users have also edited Wikipedia, resulting in a truly unprecedented base of contributors for this one community.

The dynamic of the contributions made by these individuals has been scrutinized in the past. At times, the goal has been to emphasize the egalitarianism and openness of the project. For example, Andrew Lih's (2009) insider account, *The Wikipedia revolution*, was subtitled, *How a bunch of nobodies created the world's greatest encyclopedia*, in an apparent attempt to promote the ideal of the "wisdom of the masses" over the role of either experts or organizational structures. Yet systematic research has uncovered the more complex underpinnings of Wikipedia's editorial process, such as the findings of Kittur and Kraut (2008, 2010; see also Kittur et al. 2009a, b), which highlighted the importance of coordination (operationalized as inequality assessed via the Gini coefficient) in constructing content.

Various scholars have explored the nature of Wikipedia's content production systems, with a particular emphasis on the relatively structured nature of the interaction (Brandes et al. 2009). Others have investigated the topics of roles and leadership, albeit mostly from a qualitative perspective or by focusing only on the social roles adopted in very specific contexts (Konieczny 2009). Arazy and his colleagues (Arazy et al. 2015, 2017) went beyond these narrow approaches, taking a closer look at the diversity and quasi-permanence of roles of Wikipedia. They ultimately found that roles are defined through an increasing tendency to invest oneself in the project by taking over and "curating" specific articles, thus validating the idea of a functional role in which an individual leads by active work. This volume intersects and builds upon this idea at a more abstract level of analysis, as we ask the broader question of how and why online communities generate stable elites like those that Arazy et al. observed.

Social collaboration on Wikipedia has also been researched in some depth, at times under the rubric of social network approaches, and typically with the goal of detecting communities and patterns of cooperative work (Yang et al. 2013). Finally, the evolution of Wikipedia as a social space has also been investigated using social network approaches (Capocci et al. 2006) as well as more traditional perspectives (Kittur et al. 2007). Previous work has only intermittently examined these systems from a quantitative, longitudinal perspective, however, and aside from the very recent exceptions published by Arazy et al. (2015) and Muchnik et al. (2013), they certainly have not developed any high-level abstract explanatory mechanisms for the development of social collaboration systems over time. Likewise, the other core questions in this volume—the evolving interaction between contribution inequality and leadership role stability, the mechanisms by which a contributor ascends to a leadership role, and the multidimensional nature of organizational configuration change—have not been previously addressed. This text serves to fill these gaps in the literature.

Chapter 2 is dedicated to reviewing some of the literature on social media, with special attention paid to the perspective of the studies dedicated to Wikipedia, through which we explore and clarify the social structuration and inequality that is present in social media spaces. In the remainder of part I, we continue with an analysis of social structuration and online elite resilience, which provides a key underpinning for our theoretical understanding of how social media develop their adhocratic power. Finally, in part II of this volume, we introduce, present, and employ a theoretical mechanism for future work on organizational structures that extends beyond adhocracy alone, together with some social structural analyses that serve to validate it.

References

Alexa (2016) The top 500 sites on the web. http://www.alexa.com/topsites. Accessed 14 Oct 2016

Arazy O, Nov O (2010) Determinants of Wikipedia quality: the roles of global and local contribution inequality. In: Inkpen K, Gutwin C, Tang J (eds) Proceedings of the 2010 ACM conference on computer supported cooperative work. ACM Press, New York

Arazy O, Ortega F, Nov O, Yeo L, Balila A (2015) Functional roles and career paths in Wikipedia. In: Cosley D, Forte A, Ciolfi L, McDonald D (eds) Proceedings of the 18th ACM conference on computer supported cooperative work & social computing. ACM Press, New York

Arazy O, Lifshitz-Assaf H, Nov O, Daxenberger J, Balestra M, Cheshire C (2017) On the "how" and "why" of emergent role behaviors in Wikipedia. Paper presented at the 20th conference on computer-supported cooperative work and social computing (CSCW 2017), Portland, 25 Feb–1 Mar 2017

Ayers P, Matthews C, Yates B (2008) How Wikipedia works: and how you can be a part of it. No Starch Press, San Francisco

Bales RF (1950) Interaction process analysis: a method for the study of small groups. Addison-Wesley Press, Cambridge

Bales RF (1953) The equilibrium problem in small groups. In: Parsons T, Bales RF, Shils EA (eds) Working papers in the theory of action. Free Press, New York, pp 111–161

Barabási A-L (2014) Linked: how everything is connected to everything else and what it means for business, science, and everyday life. Basic Books, New York

Benkler Y (2006) The wealth of networks: how social production transforms markets and freedom. Yale University Press, New Haven

Blau PM (1977) Inequality and heterogeneity: a primitive theory of social structure. Free Press, New York

Brafman O, Beckstrom RA (2006) The starfish and the spider: the unstoppable power of leaderless organizations. Portfolio Trade, New York

Brandes U, Kenis P, Lerner J, van Raaij D (2009) Network analysis of collaboration structure in Wikipedia. In: Quemada J, Leon G, Maarek Y, Nejdl W (eds) Proceedings of the 18th international conference on the world wide web. ACM Press, New York

Britt BC (2013) Evolution and revolution of organizational configurations on Wikipedia: a longitudinal network analysis. Dissertation, Purdue University

Bureau of Labor Statistics (2016) American Time Use Survey. http://stats.bls.gov/news.release/atus.toc.htm. Accessed 13 Oct 2016

Capocci A, Servedio VDP, Colaiori F, Buriol LS, Donato D, Leonardi S, Caldarelli G (2006) Preferential attachment in the growth of social networks: the case of Wikipedia. Phys Rev E 74(3):036116. doi:10.1103/PhysRevE.74.036116

Cision, Bates D (2009) 2009 social media & online usage study. http://www2.gwu.edu/~newsctr/10/pdfs/gw_cision_sm_study_09.pdf. Accessed 19 Feb 2015

Colón-Aguirre M, Fleming-May RA (2012) "You just type in what you are looking for": undergraduates' use of library resources vs. Wikipedia. J Acad Libr 38(6):391–399

Durkheim E (1893) De la division du travail social. Félix Alcan, Paris. English edition: Durkheim
    E (1960) The division of labor in society (trans: Simpson G). Free Press: New York
Giles J (2005) Internet encyclopaedias go head to head. Nature 438(7070):900–901
Gillmor D (2006) We the media: grassroots journalism by the people, for the people. O'Reilly
    Media, Sebastopol
Goode L (2009) Social news, citizen journalism and democracy. New Media Soc 11:1287–1305
Head AJ, Eisenberg MB (2010) How today's college students use Wikipedia for course-related
    research. First Monday 15(3). doi:10.5210/fm.v15i3.2830. http://firstmonday.org/article/
    view/2830/2476
Huberman BA (2003) The laws of the web: patterns in the ecology of information. The MIT Press,
    Cambridge
International Telecommunication Union (2016) ICT facts and figures 2016. http://www.itu.int/en/
    ITUD/Statistics/Documents/facts/ICTFactsFigures2016.pdf. Accessed 30 Jan 2017
Javanmardi S, Ganjisaffar Y, Lopes C, Baldi P (2009) User contribution and trust in Wikipedia.
    In: Proceedings of the 5th international conference on collaborative computing: Networking,
    applications and worksharing, IEEE, Washington, DC, 11–14 Nov 2009
Kittur A, Kraut RE (2008) Harnessing the wisdom of crowds in Wikipedia: quality through coordi-
    nation. In: Begole B, McDonald DW (eds) Proceedings of the ACM conference on computer-
    supported cooperative work. ACM Press, New York
Kittur A, Kraut RE (2010) Beyond Wikipedia: coordination and conflict in online production
    groups. In: Inkpen K, Gutwin C, Tang J (eds) Proceedings of the 2010 ACM conference on
    computer supported cooperative work. ACM Press, New York
Kittur A, Chi EH, Pendleton BA, Suh B, Mytkowicz T (2007) Power of the few vs. wisdom of
    the crowd: Wikipedia and the rise of the bourgeoisie. Paper presented at the 25th annual ACM
    conference on human factors in computing systems (CHI 2007), San Jose, 28 Apr-3 May 2007
Kittur A, Lee B, Kraut RE (2009a) Coordination in collective intelligence: the role of team struc-
    ture and task interdependence. In: Olsen DR Jr, Arthur RB, Hinckley K, Morris MR, Hudson
    S, Greenberg S (eds) Proceedings of the SIGCHI conference on human factors in computing
    systems (CHI 2009). ACM Press, New York
Kittur A, Pendleton B, Kraut RE (2009b) Herding the cats: the influence of groups in coordinating
    peer production. In: Riehle D, Bruckman A (eds) WikiSym 2009: proceedings of the 5th annual
    symposium on wikis and open collaboration. ACM Press, New York
Konieczny P (2009) Governance, organization, and democracy on the internet: the iron law and the
    evolution of Wikipedia. Sociol Forum 24(1):162–192
Kuk G (2006) Strategic interaction and knowledge sharing in the KDE developer mailing list.
    Manag Sci 52(7):1031–1042
Kumar R, Novak J, Tomkins A (2006) Structure and evolution of online social networks. In:
    Eliassi-Rad T, Ungar L, Craven M, Gunopulos D (eds) Proceedings of the 12th ACM SIGKDD
    international conference on knowledge discovery and data mining. ACM Press, New York
Lih A (2009) The Wikipedia revolution: how a bunch of nobodies created the world's greatest
    encyclopedia. Hyperion, New York
Matei SA, Bruno RJ (2014) Pareto's 80/20 law and social differentiation: a social entropy perspec-
    tive. Public Relat Rev 41(2):178–186
Matei SA, Bertino E, Zhu M, Liu C, Si L, Britt BC (2015) A research agenda for the study of
    entropic social structural evolution, functional roles, adhocratic leadership styles, and credibil-
    ity in online organizations and knowledge markets. In: Bertino E, Matei SA (eds) Roles, trust,
    and reputation in social media knowledge markets: theory and methods. Springer, New York,
    pp 3–33
Miller M (2012) 3 more studies examine Wikipedia's page 1 Google rankings. http://searchengine-
    watch.com/sew/study/2163432/studiesexamine-wikipedia-s-page-google-rankings. Accessed
    31 Jan 2017
Mintzberg H (1979) The structuring of organizations: a synthesis of the research. Prentice Hall,
    Englewood Cliffs

Mintzberg H (1989) Mintzberg on management: inside our strategic world of organizations. Free Press, New York

Muchnik L, Pei S, Parra LC, Reis SDS, Andrade JS Jr, Havlin S, Makse HA (2013) Origins of power-law degree distribution in the heterogeneity of human activity in social networks. Sci Rep 3:1783. doi:10.1038/srep01783

Mullan E (2009) Information inaccuracy spells trouble for user-generated websites. EContent 32(3):10–11

Newman M, Barabási A-L, Watts DJ (2006) The structure and dynamics of networks. Princeton University Press, Princeton

Nov O (2007) What motivates Wikipedians? Commun ACM 50(11):60–64

O'Sullivan MD (2009) Wikipedia: a new community of practice? Ashgate Publishing, Burlington

Ortega F, Gonzalez-Barahona JM, Robles G (2008) On the inequality of contributions to Wikipedia. In: Proceedings of the 41st Hawaii international conference on system sciences (HICSS '08), IEEE, Washington, DC, 7–10 Jan 2008

Oxford Reference (2009) A dictionary of the Internet. http://www.oxfordreference.com/view/10.1093/acref/9780199571444.001.0001/acref-9780199571444-e-50. Accessed 31 Jan 2017

Preece J, Shneiderman B (2009) The reader-to-leader framework: motivating technology-mediated social participation. AIS Trans Hum-Comput Interact 1(1):13–32

Segerberg A, Bennett WL (2011) Social media and the organization of collective action: using Twitter to explore the ecologies of two climate change protests. Commun Rev 14(3):197–215

Shaw D (2008) Wikipedia in the newsroom. Am Journal Rev 30(1):40–46

Shirky C (2008) Here comes everybody: the power of organizing without organizations. Penguin, New York

Shirky C (2010) Cognitive surplus: creativity and generosity in a connected age. Penguin, London

Silverwood-Cope S (2012) Wikipedia: page one of Google UK for 99% of searches. http://www.intelligentpositioning.com/blog/2012/02/wikipediapage-one-of-google-uk-for-99-of-searches. Accessed 20 Feb 2015

Toffler A (1970) Future shock. Random House, New York

U.S. Census Bureau (2014) Share of aggregate income received by each fifth and top 5 percent of households, all races: 1967 to 2013. http://www.census.gov/hhes/www/income/data/historical/household/2013/h02AR.xls. Accessed 11 Feb 2015

Van de Ven AH, Poole MS (1995) Explaining development and change in organizations. Acad Manag Rev 20:510–540

Waterman RH Jr (1993) Adhocracy: the power to change. W. W. Norton, New York

Wenger E (1998) Communities of practice: learning, meaning, and identity. Cambridge University Press, Cambridge

Yang J, McAuley J, Leskovec J (2013) Community detection in networks with node attributes. Paper presented at the 13th IEEE international conference on data mining, Dallas, 7–10 Dec 2013

Part I
Structural Differentiation and Social Media: Theoretical Framework

Chapter 2
Macro-Structural Perspectives on Social Differentiation and Organizational Evolution in Online Groups

2.1 Introduction

Humans are social animals. Sociability is not a mere behavioral accident. On the contrary, it is probably what makes us human in the first place. Some have claimed that language is the defining characteristic that sets the human species apart, while others claim that the distinguishing factor is reason, art, or even humor. All of these are true, to an extent, as they are key components of our ability to socialize—a behavior in which humans engage more frequently and more dynamically than any other creature.

This is, of course, when observers do not claim that evolution only incrementally pushed some human traits a bit further than other species while simultaneously restraining certain other traits. By this account, the difference between humans and other animal species would be incremental and quantitative rather than qualitative.

Regardless, sociability is the pillar upon which collaboration can occur and through which human society as we know it today was built. The vital functions of families, clubs, businesses, communities, nations, and our increasingly connected world hinge upon social interactions. One can consider the paper on which this book was printed (or the e-reader screen on which it is displayed), the money that you used to purchase it, the profession through which you secure your income, and so forth. All of these artifacts only arise from our engagement with those around us, and those essential elements of our lives in turn give rise to new means of socialization. Every element of day-to-day human existence that we take for granted as being unique to our species is, in fact, only possible because of social engagement. It should come as no surprise that *sociability* stands as the crux of *society*.

Although this volume is not intended to resolve the age-old philosophical argument about the uniqueness of human existence, it does stake out a perimeter in the debate about the nature of online sociability, which we see as a very specific phenomenon with particular characteristics. To accomplish this, we need to start with a few preliminary considerations about social behavior in general, as online sociability

© Springer International Publishing AG 2017
S.A. Matei, B.C. Britt, *Structural Differentiation in Social Media*, Lecture Notes in Social Networks, DOI 10.1007/978-3-319-64425-7_2

is a species of a given genus. This discussion is necessary because sociability is the unavoidable criterion for defining human existence and, moreover, because it can be seen as a superseding factor that may explain many other human phenomena.

In consequence, any explanation of what humans do, especially when they do something socially in a new way (e.g., invisible communities mediated by digital technologies), needs to account for the deeper meanings of the social nature of the behavior and related forms of organization. In what follows, we will outline how the human propensity for sociability shapes the emergence of some truly novel and unexpected forms of collaboration such as that observed on Wikipedia as well as other types of social media that produce useful knowledge.

2.2   Human Sociability: Possible Definitions

To begin, let us consider sociability at a rather deep and not so intuitive level. We may first state that sociability is a useful and necessary term because it does not simply state that humans are social. Rather, sociability further signals that humans are social in a special, human way. Our focus on sociability is a pointed one, as we believe that sociability encompasses the concepts of social order, social meaning, social roles, and social structure (Blau 1977). To paraphrase the famous dictum of the communication scholar Kenneth Burke, humans are "goaded by the spirit of hierarchy" (Burke 1966, p. 16). Otherwise put, there are at least theoretical reasons to believe that sociability is related to more specific concepts, such as patterned social interactions that lead to meaningful social structures and roles (Berger and Luckmann 1980). At the same time, sociability cannot be reduced to a form of transactional individualistic organization in which only individual, autonomous, locally oriented acts matter. Otherwise, much of what people do offline or online, including the staggering inequality of effort and rewards, would be hard to explain.

Further, our use and understanding of term "sociability" rest both on a deeper, more philosophical foundation and on more tangible theorems about human interaction and organization (Schutz 1967). At the deepest level, our concept of sociability taps into the deeper meanings of Aristotle's famous dictum, "*anthropos zoon politikon*." Translated by some as "the human being is a political animal," we prefer the interpretation "humans are sociable animals." In the former English rendering, the aphorism misses the true point of Aristotle's vision of human essence. His use of the word "politikon" was forced by the nature of Greek thinking at the time, which identified living in a city (polis) with living in a polity, that is, the Greek city state, which to him meant "living in a society." Yet, what Aristotle truly meant to say is not that "man" (as he understood it in a gendered way) is limited to those who lived in Greek states. His intention was to define universal man, Greek or barbarian:

> Hence it is evident that the state is a creation of nature, and that man is by nature a political animal. And he who by nature and not by mere accident is without a state, is either a bad man or above humanity; he is like the "Tribeless, lawless, heartless one," whom Homer denounces—the natural outcast is forthwith a lover of war; he may be compared to an

isolated piece at draughts. Now, that man is more of a political animal than bees or any other gregarious animals is evident. Nature, as we often say, makes nothing in vain, and man is the only animal whom she has endowed with the gift of speech. And whereas mere voice is but an indication of pleasure or pain, and is therefore found in other animals (for their nature attains to the perception of pleasure and pain and the intimation of them to one another, and no further), the power of speech is intended to set forth the expedient and inexpedient, and therefore likewise the just and the unjust. And it is a characteristic of man that he alone has any sense of good and evil, of just and unjust, and the like, and the association of living beings who have this sense makes a family and a state. (Aristotle 350 B.C.)

In Aristotle's view, the human being is unique because to function, he or she needs a community of beings like him or her. This is necessary not merely to survive in the most immediate sense, but more importantly, to grow, to become autonomous, and to contribute back to the survival of the larger community that makes his or her own subsistence possible. To this end, the community will, by necessity, use language to instruct, command, educate, and support. The social ties that the community needs to bind and support individuals to maturity will be articulated by logical thought and reasoned principles but also by emotionally expressed connections. Logic and reason and affect and emotion always occur in the context of language and communication. Communication is symbolic, and thus it conveys messages through a variety of means, from commonplace verbal discourse to highly stylized artifacts and representations of shared meaning. Such methods include both sciences and the arts, from the fine (painting, sculpture) to the performative (poetry, theatre, dance).

Throughout the process of making humans what they are through communication and culture, sociability will bestow on some certain roles, complete with associated rights, privileges, and obligations, while to others various alternative roles. Fathers and mothers rear and teach sons and daughters some basic elements of being human. These include customary habits of behavior and thought, elementary moral principles, and rules of social conduct ranging from good manners to religious and ethical principles. After a certain age, teachers take over this educational role, and eventually, humans learn their trade of being human "on the job" as members of various institutions, groups, and social arrangements that are more or less formally defined. In each of these situations, starting with what was initially defined by the family context, certain divisions and situations—at some times, of labor, and at others, of power and privilege—teach humans to work with, to submit to, or to take control over others. These constraints teach growing individuals that human affairs are patterned interactions by which individuals assume one or more roles that give them just as much as they take in terms of autonomy.

Of course, this learning on the job, "learning by doing," does not mean that the social institutions, the roles, the power arrangements, or the difference between the "haves" and "have nots" that they often entail are natural or justified. All roles and the privileges or privations they entail are the product, after all, of a place and a time in which human ideas of worth, power, and representativity are influenced by a variety of factors. These may be merit-related, or they may be mere accidents of human inequality or chance. Roles and rules are subsequently consolidated by religious or quasi-religious justifications of sacred vs. ordinary vocations or reinforced by material interest, birthright, raw power differences, or pseudo-moral principles of the "first come, first served" kind (Weber 1947).

Yet across, above, and beyond the particular ways in which roles and social forms of organization exist, and superseding any discussion about one's ethical, moral, and justifiable existence, being human inevitably hinges upon assuming and enacting roles within social institutions (Schutz and Luckmann 1989). In fact, if we are to believe that there is a possibility to improve the human species in the social and moral realm, we have to believe that this can only be done by making sure that roles are acquired or rights and obligations distributed in an equitable, just, and moral way. The alternative view that the human vocation is to free humans from all roles and organization, that human autonomy is the only measure of all good things, and that roles, rules, and obligations must be abolished has remained so far the province of utopian projects that have failed and continue to fail in practice, from the short-lived radical Anabaptist communities of the Protestant Reformation to the doomed province of some communist experiments shortly after 1917 and the hippie communes of the 1960s.

With all of this said, Aristotle's dictum is not merely another metaphoric utterance of an ancient figure about the out-of-fashion idea of "man," with all the baggage this interpretation might mean. Although he lived in a world that is quite different from ours in mores, technology, and social institutions, not to mention the implicit definition of "men" as male, Aristotle's definition of humanity nevertheless includes a formal and comprehensive idea of sociability, which by necessity and definition includes communication, symbols, meaning, reason, and emotion. This urges us to highlight the human vocation as engaging with others in a meaningful, patterned way, emphasizing sociability in particular, as the main way to understand humans.

Returning full circle, human sociability is thus the crux of human existence in that its regular patterns of interaction, guaranteed by norms and role and reinforced by values and symbols, are, in the end, the core of human life. This philosophical principle, however, cannot simply be proclaimed in isolation. In stating it, we need to both fall back on and project a set of theorems about human interactions and human social life. These ultimately suggest that regular, role-based interactions with others are the *sine qua non*—the indispensable ingredient—not just of a *good* human existence but of *all* human existence.

2.3 Sociability and Structural Differentiation

Classical sociology (Alexander 1982; Levine 1995; Nisbet 1993), which has been extended through rigorous theoretical and practical work in modern social psychology (Alexander et al. 1987; Hare et al. 1965; Hogg 1992), emphasized the role of social differentiation and the emergence of division of labor as essential instruments for understanding the evolution of human sociability. From Durkheim (1893) to

Bales and Slater (1956), and further through a variety of social scientific studies, sociology has delineated a theory of social organization that strongly relies on the importance of social roles and social structures for understanding human sociability. A simpler way to put it is that sociability, as a broad philosophical concept, relies on a conceptual infrastructure. This has, at its core, the idea of social structure: patterned, predictable interactions that emerge from a given division of labor. Here, division of labor expands past the narrow concept of material production, entailing any and all human collaborative processes that lead to a finished product, material or immaterial (Friedson 1976; Merton 1934).

Any division of labor necessarily implies a process of structural differentiation in which certain group members perform certain activities unique to their roles (Friedson 1976). The emergence of this role and performative allocation is not merely conditioned by formal arrangements; it stems from the natural preconditions of effective group work itself. One of these preconditions is that the amount of work expended by each individual to monitor other members should be minimized, while the amount of work performed that directly contributes to the final production of the intended object of the collaboration must be maximized. Groups in which roles are poorly differentiated force members to spend a significant amount of time monitoring other members in order to prevent replication of duties or work performed, to learn new processes and norms, or to avoid mistakes. In the absence of well-defined and distinct roles, individuals also have to ensure that costs and benefits are equitably distributed. As groups increase in size, the amount of information that needs to be collected about the other members of the group and about the state of the group as a whole increases. Collection and processing time expands. Soon, individuals spend more and more time monitoring each other, devoting less and less time to working on the task at hand. Thus, a group of undifferentiated collaborators will eventually collapse under the weight of its communication and coordination demands.

However, as soon as a subset of individuals begins to specialize in communication and coordination—in other words, as soon as at least two roles emerge, that of coordinator and "worker"—groups can again grow without putting undue strain on the collaborative process, at least up to a point. Eventually, when the coordinators themselves receive so much information that it exceeds their physical processing capabilities (natural or augmented by various technologies), the group will again stop growing and its efficiency will start to decay. A new round of structural differentiation, by which coordinators specialize in specific functions—some collecting, some processing, and some relaying information—can unlock the next horizon of organizational growth. Similarly, as workers themselves start to specialize in specific tasks, which are coordinated by dedicated role-playing individuals, the organization will begin increasing in sophistication and improving its ability to grow and explore new levels of complexity.

Of course, growth has its limits, even when supported by structural differentiation (Blau 1972). As new layers of command, control, and coordination emerge, the roles assigned to these functions are connected to the actual work through increasingly extensive and longer chains of command (Blau and Schoenherr 1971). Roles themselves become ever more autonomous, gaining new identities and missions that are sometimes disconnected from the job at hand. In short, especially complex organizations with multiple layers of command and control tend to place the mission of mere organizational preservation and the interests of the roles found at the top of the pyramid ahead of the real mission of the organization. Furthermore, as information is collected by some roles, processed by others, and disseminated by yet another role-playing group, signals become attenuated by noise. Information is inevitably lost along the way or, worse yet, cherry-picked in order to support individual needs or political ends rather than the goals of the organization (Blau 1970).

Therefore, structural differentiation is not necessarily an absolute, normative solution to the requirements of human organization, nor it is infinitely expandable (Blau 1970). Yet, for organizations that emerge spontaneously, the path of structural differentiation is the one that they follow up to a certain point. We should expect it to emerge in most situations where human groups aim to solve a certain task, as role-based differentiation provides context for individual work and life.

The debate mentioned so far not only follows in the footsteps of classical sociology harkening back to Durkheim, but it is also connected to more recent debates about the natural limits of human organization as facilitated by normed roles and interaction patterns (Burgers et al. 2009; Cullen et al. 1986; Mayhew et al. 1972). It also intersects with Olson's (1971) ideas about the prerequisites of "collective" action and the "free riding problem." Just like in Olson's public goods production processes, in our research we noticed that online collaboration works well when there are selective incentives for the active members. In our case, such incentives are a sense of ownership and the intrinsic psychological reward active users get from shaping a given online collaborative project. These motivators lead to a production system that is driven by a small group of contributors, who have both a higher level of investment in and reward from the project, while the rest more or less benefit from their work. Our work also intersects with Marwell and Oliver's (2007) ideas on critical mass, which established that a critical mass of dedicated individuals may propel social movements and voluntary projects. Like them, we think of this group of dedicated individuals as constitutive. We also agree that the critical mass concept should not be seen as the mere minimum number of participants, but as the minimum number of *active* (high contributing) participants needed to jump start a project. In other words, the "1% effect" that underlies our argument (the top 1% of users effectively shape and are shaped by online collaboration through social differentiation) is closely related to the critical mass argument.

However, our work is not a mere derivation of Olson's or Marwell and Oliver's prior work. We are not directly interested in the public goods nature of the products generated by online groups, as Olson would be, nor in the "production functions" that may or may not moderate the self-sustaining chaining-out process of collaboration, which are at the heart of Marwell and Oliver's work. Instead, we focus on the social and communicative dynamics that lead in time to social structuration and differentiation. While complementary, this is a distinct problem, with its own research questions.

Our work also intersects with other scholarly debates, especially those that emerged around the discussion introduced by Coase (1937) about the nature and limits of formal business organizations (Ellickson 1989; Gibbons 1999), as well as the debate about the possibility of "peer production" (Benkler 2002).

While these are fascinating topics, our book will not formally or directly engage such arguments. Our focus is much more specific. Simply put, we aim to test if voluntary knowledge production organizations such as Wikipedia can be seen through a structural differentiation perspective, if specific phases can be discerned in their life cycles, and if elites emerge while differentiation is ongoing. We also investigate the process of organizational change more broadly, proposing a new multidimensional perspective for understanding organizational evolution and the "motors" that move it forward.

2.4   Online Sociability and Structural Differentiation: Connections and Directions of Study

To conclude this overview chapter, it is worthwhile to outline in summary the theoretical ideas behind our specific structural differentiation theoretical framework and the possible ways in which it can explain knowledge production organizations online. In particular, the following pages focus on some core concepts, especially sociability, differentiation, and adhocracy.

Since this volume is not one of general theoretical sociology, but one rooted in communication research and dedicated to examining specific social groups mediated by technology, sociability needs to be adapted to serve the more immediate research context. In that respect, online sociability is the set of generalizable rules by which we can describe regularities in online social interactions and communication. Here, again, we do not take sociability to be an empty tautology by which we simply designate what people do in online groups (viz., interacting with other people online). To be useful, online sociability needs to account for several factors: low barriers of entry and exit, weak pressures to conform and commit, and voluntary and natural interactions (Blanchard and Markus 2004; Ciffolilli 2003; Rheingold 2000).

Online sociability should thus be seen as structured and flexible, hierarchical and relatively decentralized, normed and anti-normative, and authoritative while not authoritarian. Of course, this only offers a general set of tendencies, thus allowing online sociability to take a range of possible forms. Yet, through embracing these attributes, online sociability comes to resemble other, more traditional terms, such as "adhocracy." First proposed by Warren Bennis (1968) and further developed by Alvin Toffler (1970) and Henry Mintzberg (1979), among others, adhocracy describes a form of social organization that supports organically emergent groups in which roles are achieved, not prescribed, and in which members enter and leave the organization at a certain rate, which is neither very high nor very low. At the same time, adhocracy does allow for roles which, even if temporary, are occupied by certain members for a nontrivial period of time. Such roles carry real power, and this power has consequences. Similarly, control and communication mechanisms exist alongside norms, rules, and enforcement mechanisms, thereby enabling the development and application of a power structure based in functional roles. Yet, again, adhocratic roles are weakly formed and achieved by completing their attendant obligations, not by formal induction, nomination, or election.

While it is not a new term, we aim to instill adhocracy with sufficient theoretical rigor, buttressed with empirical evidence, to justify a new take on an old problem: how and within what parameters do new forms of human organization appear?

Within this context, a significant theoretical effort should be made to better circumscribe the role that elites and leadership groups play in the social context of adhocracy. Here, the most important issue is that of articulating the contributions that elite groups offer in structuring social groups online. This point is developed in the next chapter.

References

Alexander JC (1982) Theoretical logic in sociology: the antinomies of classical thought: Marx and Durkheim. Routledge, New York

Alexander JC, Giesen B, Münch R, Smelser NJ (1987) The micro-macro link. University of California Press, Berkeley

Aristotle (350 B.C.) Politics: book I. English edition: Aristotle (1984) Politics: book I (trans: Jowett B). Princeton University Press, Princeton

Bales RF, Slater PE (1956) Role differentiation in small decision-making groups. In: Parsons T, Bales RF (eds) Family, socialization and interaction process. Routledge, London, pp 259–306

Benkler Y (2002) Coase's penguin, or, Linux and the nature of the firm. Yale Law J 112:369–446

Bennis WB (1968) The temporary society. Harper & Row, New York

Berger PL, Luckmann T (1980) The social construction of reality: a treatise in the sociology of knowledge, 1st Irvington edn. Irvington Publishers, New York

Blanchard AL, Markus ML (2004) The experienced "sense" of a virtual community: characteristics and processes. ACM SIGMIS Database 35(1):64–79

Blau PM (1970) A formal theory of differentiation in organizations. Am Sociol Rev 35:201–218

Blau PM (1972) Interdependence and hierarchy in organizations. Soc Sci Res 1(1):1–24

Blau PM (1977) Inequality and heterogeneity: a primitive theory of social structure. Free Press, New York

Blau PM, Schoenherr RA (1971) The structure of organizations. Basic Books, New York

Burgers JH, Jansen JJP, Van den Bosch FAJ, Volberda HW (2009) Structural differentiation and corporate venturing: the moderating role of formal and informal integration mechanisms. J Bus Ventur 24(3):206–220

Burke K (1966) Language as symbolic action: essays on life, literature, and method. University of California Press, Berkeley

Ciffolilli A (2003) Phantom authority, self-selective recruitment and retention of members in virtual communities: the case of Wikipedia. First Monday 8(12). doi:10.5210/fm.v8i12.1108. http://firstmonday.org/article/view/1108/1028

Coase RH (1937) The nature of the firm. Econ-New Ser 4(16):386–405

Cullen JB, Anderson KS, Baker DD (1986) Blau's theory of structural differentiation revisited: a theory of structural change or scale? Acad Manag J 29(2):203–229

Durkheim E (1893) De la division du travail social. Félix Alcan, Paris. English edition: Durkheim E (1960) The division of labor in society (trans: Simpson G). Free Press, New York

Ellickson RC (1989) The case for Coase and against Coaseanism. Yale Law J 99:611–631

Freidson E (1976) The division of labor as social interaction. Soc Probl 23(3):304–313

Gibbons R (1999) Taking Coase seriously. Adm Sci Q 44(1):145–157

Hare AP, Borgatta EF, Bales RF (eds) (1965) Small groups: studies in social interaction. Knopf, New York

Hogg MA (1992) The social psychology of group cohesiveness: from attraction to social identity. Harvester Wheatsheaf, Hemel Hempstead

Levine DN (1995) Visions of the sociological tradition. University of Chicago Press, Chicago

Marwell G, Oliver P (2007) The critical mass in collective action, Reprint edn. Cambridge University Press, Cambridge

Mayhew BH, Levinger RL, McPherson JM, James TF (1972) System size and structural differentiation in formal organizations: a baseline generator for two major theoretical propositions. Am Sociol Rev 37:629–633

Merton RK (1934) Durkheim's division of labor in society. Am J Sociol 40(3):319–328

Mintzberg H (1979) The structuring of organizations: a synthesis of the research. Prentice Hall, Englewood Cliffs

Nisbet RA (1993) The sociological tradition. Transaction Publishers, New Brunswick

Olson M (1971) The logic of collective action: public goods and the theory of groups, second printing with new preface and appendix, Revised edn. Harvard University Press, Cambridge, MA

Rheingold H (2000) The virtual community: homesteading on the electronic frontier. MIT Press, Cambridge

Schutz A (1967) The phenomenology of the social world. Northwestern University Press, Evanston

Schutz A, Luckmann T (1989) The structures of the life world. Northwestern University Press, Evanston

Toffler A (1970) Future shock. Random House, New York

Weber M (1947) The theory of social and economic organization. Oxford University Press, New York

Chapter 3
Specifying a Wikipedia-Centric Explanatory Model for Online Group Evolution and Structural Differentiation

3.1 Introduction

In the previous chapter, we set up the broader sociological underpinnings of our argument. This chapter will focus on the more immediate and tangible mechanisms that shape the emergence and evolution of social media groups, such as those that build Wikipedia. Our theoretical argument will devote particular attention to the role played by contributing elites in organizing and sustaining collaborative activity.

It is worth noting that the mechanisms described here may transcend specific social media, and likewise, there might exist some specific features of particular media that do not entirely intersect with our more generalizable propositions. Before dealing with these issues, however, one important premise should be declared immediately. Some might wonder why Wikipedia is considered a collaborative social medium. One could, after all, imagine a knowledge-construction community of this sort operating more like a factory, in which work proceeds in the manner of an assembly line with minimal interaction between contributors. While some work on Wikipedia is indeed completed in isolation, the overall production process is collaborative and intensely interactive. Simply put, Wikipedia is an example of collaborative social media because its content is contributed by users who engage in collaborative practices that extend over long periods of time. Users freely engage with one another's work when editing articles, and they also directly interact with one another via open discourse on discussion pages. This collaborative process is a prime example of sociability, so Wikipedia must itself be considered a social medium.

With that in mind, this chapter provides answers to the following questions: How, exactly, do social media groups like Wikipedia come together? Through what distinct phases do they progress, if any? Do they create structures and roles for their members to fill? How does this matter?

© Springer International Publishing AG 2017   31
S.A. Matei, B.C. Britt, *Structural Differentiation in Social Media*, Lecture Notes
in Social Networks, DOI 10.1007/978-3-319-64425-7_3

3.2   Volunteering and Social Inequality

Let us start by stating that collaborative online social media groups are, for the most part, a species within a broader genre. The larger sociological category in which they may be positioned is that of voluntary organizations. These have been present in everyday life for a long time, from loosely defined club affiliations to highly structured volunteer service organizations such as the famous Lions, Shriners, or Kiwanis. Some of the assumptions we currently utilize to explain this larger class of organizations should also apply, to a certain extent, to collaborative social media.

Yet at the same time, collaborative social media groups are task-oriented in the narrowest sense. Each such group deals with one objective at a time: build a Wikipedia article, maintain a conversation or dispute on Reddit, make content viral on Twitter, and so forth. Because of this, social media groups also resemble communities of practice (Wenger 1998). Furthermore, online groups, especially when they are small, may take a variety of other forms through their functional requirements, including that of an entrepreneurial or bureaucratic organization. We will expand on this later, especially in Chap. 8.

Whenever individuals join forces to perform a voluntary task, the groups they form ultimately split into at least two categories of members: leaders and followers (Shafiq et al. 2013). The working mom who takes it upon herself to organize a neighborhood watch group, the dad who coaches the local little league baseball team, the girl who is the president of the school robotics club, and the boy who clocks the most hours volunteering at the local library are all well-known figures in local communities that foster volunteer organizations. They stand out not only by personal involvement but also by the amount of time and effort they commit to their chosen activities.

To make the point clearer, recall that as we noted in Chap. 2, between 2003 and 2011, the top 20% of all volunteering individuals contributed 50% of the time volunteered in the USA, as calculated from raw data retrieved from the American Time Use Survey (Bureau of Labor Statistics 2016). In other words, if you compare the top 20% of volunteers against the rest, the average individual in this elite group contributed four times as many hours as the average person in the bottom 80%.

This is a significant difference in workload, to be sure. Given this prevalence of unequal involvement in volunteer activities in general, it should come as no surprise that contributions are also unequal for one of the most gigantic volunteer collaboration projects in the world, Wikipedia. Between 2001 and 2010 (the timeframe of our retrospective study), there were 235,701,162 edits made by 22,792,847 unique users. 19,680,637 of these accounts consisted of IP addresses attributed to anonymous individuals, while the other 3,112,210 were user-created accounts. Logged-in users, which represent only 16% of all users, contributed almost twice as many edits as their anonymous peers—they were responsible of almost two-thirds (68%) of all revisions made to Wikipedia, a clear majority of the workload. Further, if we analyze contributions not by edits, but by the amount of work estimated by a measure of effort,[1] observed inequality increases dramatically: the top 1% logged-in user accounts are responsible for 82% of the effort of all logged-in users. More importantly, 77% of the effort is produced by the top 1% of users, regardless if they were logged in or not.

In brief, when people volunteer their time and effort to work on a group project, be it online or offline, they do so at different rates. With that in mind, what explains the gap between high-contributing and low-contributing individuals? We propose that such inequality is, among other things, a marker of social organization.

3.3 Why Inequality Matters

At the heart of our book lies a simple proposition: the often-observed inequality found in everyday interactions online (Barabási 2014; Nielsen 2006; Sauermann and Franzoni 2015) is not an accident nor do its origins lie in factors extraneous to the online social dynamics. Thus, inequality is neither a triviality to be dismissed nor a side effect to be decried. Inequality online is, in fact, closely related to other types of collaborative inequalities, many of which are found offline. When compared with the longstanding Pareto principle (80% of something is produced by 20% of the potential sources), we immediately recognize that online inequalities are not very different from the uneven distributions of effort, contributions, rewards, or inputs in everyday life (Newman 2005; Pareto 1906). While the proportions differ from one situation to the next, and although online groups tend to feature a noticeably higher skewness (often more like 90/9/1; see Nielsen 2006), the observed inequality is itself not particularly unusual.

As this phenomenon may, at first glance, appear to be an inconsequential statistical artifact, it demands a parsimonious interpretation that can shed light on its significance. We propose that observed inequalities in effort distributions are, at least in part, a reflection of the quasi-natural and self-regulatory process of structural differentiation in human organizations (Blau 1970).

The relationship between inequality and structural differentiation can be established at several levels. First, as discussed in the previous chapter, larger groups impose higher coordination and control costs on individual members if they are all to perform their respective activities by themselves, in isolation. As soon as individuals assume differentiated roles, some of them focusing more on communication and coordination or on setting expectations for the general effort, the group structures itself and work distribution becomes more efficient.

Furthermore, it is not even essential to have a subgroup of members engage in explicit coordination and control activities. It is sufficient for some individuals to work more than their peers. As a comparatively larger amount of effective and useful work is done by fewer individuals, the group of contributors who matter shrinks, and with it, the need for coordination and control also declines, as most work is

---

[1] See Chap. 5 and Appendix A for a definition and measurement of effort in this context.

done within a rather narrow circle. In fact, this is probably what we witness on Wikipedia and other large-scale online production systems in which effective collaborative processes are maintained over time with minimal explicit coordination or control—there is little need for extensive mechanisms to serve these purposes if the circle of significant contributors is small.

Given these assumptions, we should acknowledge that the process of peer production, as it is called in some quarters, is a complex one whose understanding demands multiple analytic perspectives. The summary offered by Benkler et al. (2015) provides a good overview of the problem space. One of the core conclusions of this overview is that peer production has drifted at least conceptually, if not in actual practice, away from a vision of locally coordinated, horizontally distributed tasks. This notion has been supplanted by a more complex and realistic explanatory framework that takes into account the fact that leaders in collaborative spaces lead by doing more and by staking out specific territories, which they *de facto*, if not *de jure*, take into ownership (Arazy et al. 2015).

Much of the research on collaborative inequality has been dedicated not only to theoretical explanations of online interaction and collaboration phenomena, but also to their practical implications, especially those related to growth and recruitment (Halfaker et al. 2012). The concerns appear to be legitimate, as some sites, including Wikipedia, have seen slower growth in recent years. Yet, the worries might not be so legitimate if we consider that online social spaces may follow an evolutionary process with an upper limit for growth. Expectations of linear growth are not always warranted, especially after social groups go through a period of structural differentiation, eventually reaching a plateau of patterned interaction and role allocation. As we suggest below, interaction and inequality on Wikipedia actually reached this level quite soon after launch, so it has therefore been in a state of quasi-equilibrium for quite some time, especially in terms of the structural differentiation and contribution inequality within the collaborative effort.

3.4   Social Roles and Structural Differentiation

Speaking about inequality, how should we translate it in a more direct and intuitive manner using the language of structural differentiation? At its very core, the idea suggests that as voluntary groups grow, some members tend to assume leading positions based on the amount and type of work they perform (Preece and Shneiderman 2009). This is nothing new, as Welser et al. (2011) as well as, more recently, Arazy et al. (2015) proposed specific sets of roles defined by substantive, technical, and communication attributes. While these roles involve both qualitative (what) and quantitative (how much) descriptions of the work being done, in what follows, we will focus on the quantitative differentiation of leaders from the rest of the group.

From our perspective, leaders are those who lead by example (Plowman et al. 2007). By doing so, leaders implicitly become anchor points for everybody else's work. A recent, path-breaking study by Muchnik and collaborators showed quite convincingly that there is a direct connection between ranking in the "head" of a power-law distribution of effort on Wikipedia and the extensiveness of one's work and communicative connections (Muchnik et al. 2013). The probability that an ordinary, less involved collaborator will touch upon a top contributor's work increases quite substantially as collaborative inequality increases. Consequently, top contributors act as driving forces that shape the mores, norms, and standards of the collaborative community. Furthermore, leaders stake out areas of interest, which they then "shepherd" (Halfaker et al. 2009). This personal stake in their respective slices of the community and the resulting product provides top contributors with an extra motivation to contribute, which further accentuates the uneven distribution of effort while giving the leaders the feeling of literally being in control of the process (Panciera et al. 2009).

The net outcome of leading by doing more is the tendency of groups to self-organize. This organization includes both formal and informal roles (Blau 1977). On Wikipedia, editors are often given badges or "barnstars" to signify their leading positions on the project (Kriplean et al. 2008). These signs of recognition along with subtler interactional cues, such as participation in discussions on "talk" pages,[2] may also help a user become a formal "admin" through election by his or her peers. Such users have the power to close certain articles to editing, for example, or to participate in procedures that may even lead to banning other users. Thus, in some cases, the unspoken, informal leadership roles that members adopt are eventually converted into formal leadership positions.

Yet—and this is a critical caveat—informal roles, which we also call "functional roles" (Arazy et al. 2015), are the defining social roles that make Wikipedia and other online voluntary organizations work. These informal roles are carved out by doing and by working more and harder than everybody else. Such roles inform the formal administrative hierarchy and are, in fact, a precondition for official advancement. As such, the informal roles are those that tend to matter most.

In previous work (Matei et al. 2015), we have developed this argument more formally for Wikipedia. In the text that follows, we reiterate some of those ideas as scaffolding for the present, broader investigation, which more strongly emphasizes the role of structural differentiation in the evolutionary process of social media spaces. For more information on the growing literature on Wikipedia, see Jullien (2012).

In the past, collaborative inequality was explained through individual motivations and abilities (Javanmardi et al. 2011; Nov 2007). In more recent work, however, Welser et al. (2011) and Arazy and Nov (2010) suggested that this inequality might also reflect a tendency toward role specialization. We pursue this insight further, although via our own structural differentiation perspective, which is not explicitly present in the previous literature. In this respect, we tend closer to the theoretical arguments of Blau (1970), Mintzberg (1979), and Bales and Slater (1956). The latter is particularly important, as Bales' research on small groups showed that the

---

[2] On Wikipedia, each article is accompanied by a "talk" page, a space where the authors or readers can discuss the editorial process.

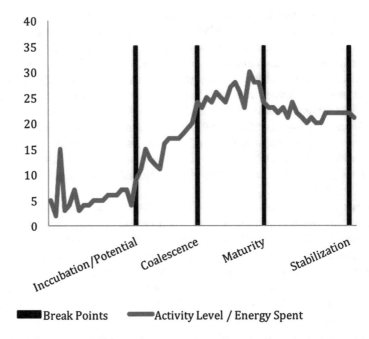

Fig. 3.1 The four stages of evolution for communities of practice (hypothetical example)

emergence of functional leaders—those who lead by doing—is a constant process in all spontaneous organizations.

These early insights are cross-pollinated with Wenger's (1998) proposition that organically emerging organizations, such as communities of practice that are created by free association, follow an evolutionary path that can be divided into distinct phases. Communities of practice in particular, and voluntary organizations with explicit collaborative goals more generally, typically emerge through a four-stage process (Fig. 3.1).

The first stage is that of incubation or gathering potential. Participants discover their own abilities, resources, skills, and capacity to collaborate, as well as those of the group itself. This is a period of numerous and, at times, momentous shifts in technologies, rules, strategies, commitment, amount of time spent on the project, and member-to-member interactions. The distribution of effort may likewise shift quickly and dramatically, from relatively even to extremely uneven.

In the second phase, coalescence, participants start orienting toward each other, with some individuals assuming increasingly large shares of effort. The collaboration develops such that it can be defined as a space that encompasses clearly distinguishable "in" (high-effort) and "out" (lower commitment) members.

The third phase is that of maturing. At this point, interactions between members and expectations of effort, especially those of longtime members, consolidate. Collaboration becomes routinized, with high-effort members contributing much of the content and consequently shaping the overall collaborative process and product.

Finally, there is the fourth phase, stabilization, during which the leaders take a step back. This is also known as the "stewardship" period. While the leaders still dominate the work effort itself by almost an order of magnitude, they also dedicate some of their time to other tasks, including control, coordination, and adjudication of conflicts. The change of focus allows the other members of the project to increase, in relative and limited terms, their share of the effort. The stabilization period is, as the label suggests, one in which we may observe a relatively stable, albeit skewed, level in distribution of effort. Once matured, the group maintains the distribution of effort between the top contributors and everybody else at a highly skewed level, which replicates itself in successive periods. This is the phase in which we may say that structural differentiation has reached its maximum potential.

Notably, most communities of practice and voluntary associations do not persist in the "stabilization" phase forever. Cohort or cultural changes introduced by social contact with other projects or induced by social context pressures may draw some members away from the project or may change the characteristics and goals of the members. When this happens, some organizations enter a period of dramatic transformation that results in either organizational identity change or splintering. Importantly, those organizations that resist change or otherwise fail to transform upon exiting the stabilization phase instead slowly decline and die.

The phase approach to organizational evolution and growth is extremely important for detecting changes in the structural differentiation process.[3] The reverse is likewise true, as any detectable changes in structural differentiation may indicate that the group has entered a new phase of activity. The main challenge is to identify a measure that meaningfully aligns changes in distribution of effort or contributions with claims of structural differentiation. We achieve this by using social entropy as a core measure, as we will discuss below and in later chapters.

The goal of uncovering phases in activity over time demands a method to detect boundaries in the evolutionary process (Cummings and Worley 2014; Lewin 1947). These are points of inflection in the trajectory of the measurable indicator of uneven contributions (or, perhaps, alternative indicators representing other collaborative dynamics, as will be discussed in Chap. 7). Contribution unevenness is interpreted as a sign of social structuration that may be expressed as social entropy (O'Connor 1991), whereas an evolutionary boundary is defined as a statistically measurable shift in entropy (Leydesdorff 2002).[4] Such boundaries are indicators of social movement toward or away from social structuration. As we trace the evolution of structural differentiation and its discrete phases, we effectively map the "embryology" of the collaborative space. This is a core contribution of this volume and of the research agenda that made it possible.

---

[3] We borrow the idea of phased development mostly from Wegner, as indicated above. Yet, there are other views, such as those of Tushman and Romanelli (2008), which can also be taken into account. Our view, in distinction to Tushman and Romanelli, who proposed a punctuated equilibrium perspective, takes a more incremental approach to phase transitions.

[4] The cited paper refers to informal scientific groups, which is highly relevant for this context.

With all of that said, a mere change in the entropy indicator is not sufficient reason to make the claim that an evolutionary phase has started or ended. Detectable changes in social differentiation and social aggregation via measurements of contribution inequality need to be validated through a second, dependent measure, which demonstrates that any observed inequality is the product of a quasi-stable elite group consistently contributing more than the rest of the community, rather than an artifact merely resulting from members taking turns in making contributions and rotating into and out of arbitrarily defined "elite" positions (Dellarocas et al. 2014).

To put it another way, we are keenly aware of the possibility that a group that is uneven in contributions can be open or closed. The roster of top contributors may have a constant composition over time, with a particular group of individuals perpetually participating and occupying leadership positions. Alternatively, it may instead be that the "top contributors" are continuously refreshed with new recruits such that a wide range of members periodically ascend into and descend from the highest ranks, as in a game of "musical chairs."

In the former scenario, the group is closed. In the latter, it is open. When an open group displays high contribution inequality, this can simply reflect the fact that not all members can sustain high levels of contribution all the time. Members of the group may distribute their efforts in time, with some of them working more during certain intervals and less during others. Thus, while during any given interval it would appear that the group is dramatically divided into high- and low-productivity members, contributions over time may, in fact, even out.

However, if we concurrently observe uneven processes and "elite stickiness," wherein members of the elite group tend to remain in an elite role for a prolonged period, and if we can determine that unevenness precedes this observed stickiness, we may conclude that the phases and phase shifts in the collaborative unevenness measure (entropy) are indeed the product of structural differentiation.

A key element of this type of analysis, and the associated theories that drive it, is the specific manner in which we conceptualize social structuration and structural differentiation. In Chap. 4, we elaborate on the idea of structural differentiation as a non-entropic phenomenon. The reasons and articulations of this concept are presented there in more detail. For the sake of immediate clarity and the continuity of this argument, however, it is at least worth reiterating that our argument rests, in large part, on the assumption that uneven distributions can be characterized by a social entropy measure which indicates whether or not the uneven collaborative process is evidence of a meaningful social structure.

In this respect, we take our cue from Shannon and Weaver (1948), who proposed an entropy-derived measure to characterize a process that is intrinsic to human phenomena, namely, communication. Shannon and Weaver made the claim, which was later substantiated, that an even occurrence of symbols in a communicative system betrays a state of randomness and entropy. Such systems, when observed, are inherently disorganized, noisy, and lacking meaningful signals. In such systems information is minimal or absent. At the same time, when certain symbols of a communication system occur more often than others, beyond a rate that chance alone would predict, meaningful information and signals are more likely to be present. In this state, the system as a whole is more likely to be "structured." More recent literature, including trade publications, revived this insight, expanding it to a more comprehensive view of reality as an information-based system (Seife 2007).

We take this insight to the next level (Bailey 1990), applying the logic of this communication-based analysis to social phenomena in the manner suggested by Osgood and Wilson (1961). Social systems, which are similar to and overlap with communication systems, are comprised of elements (individuals) that are in various states of presence within a range of contexts: collaborative, communicative, interactive, persuasive, and so forth. When all individuals are in the same state of presence (i.e., they collaborate in the same proportion, talk the same amount, interact at the same rate, etc.), the social system is more or less random and disorganized, as everybody is doing the same things at the same rate and no leaders are emerging to guide any component of the interaction. On the other hand, when some individuals are more present than others, whether through communication or other types of behavior, the system starts displaying signs of organization. This insight is not speculative. It relies on well-established theoretical ideas within communication and sociology (Backstrom et al. 2006; Georgescu-Rogen 1971; Schramm 1955).

In this context, Osgood and Wilson's (1961) insights are particularly germane. In their groundbreaking but almost forgotten text, *Some terms and associated measures for talking about communication*, they argue that the degree of structuration or organization of a communicative and social system is the inverse of its level of uncertainty. According to them, "uncertainty is a characteristic of a system which increases with the number of its states and the degree to which these states occur with equal frequency or probability" (2). Uncertainty maps onto entropy, and in this respect, even or random states are, in effect, uncertain states.

At the same time, we must emphasize that while uneven distributions suggest the presence of social structures, the structures themselves might not be static, monolithic social orders. Individuals who are at the top of the functional hierarchy are not necessarily their rulers. Neither do they wield absolute directive control over the community. Leadership is, again, defined in terms of what one does. One's role as a leader lasts only as long as the individual works harder than everybody else. Likewise, competing leaders, with competing roles, might exist.

When taking all of this into account, we can anticipate a certain amount of leadership turnover. In this text, we report our measurements of this turnover, which we use to more accurately assess the level of relative elite resilience ("stickiness") and to discern whether it represents a form of leadership that could be called "adhocratic" or "just-in-time leadership." Adhocracy, as a term, was first popularized to the public by Toffler (1970), and Mintzberg (1979), Van de Ven and Poole (1995), and Waterman (1993) directly and indirectly refined it through their explicit and implicit intellectual theorizing about organizational structures. For now, we may defer to a more concise definition offered by the Oxford *Dictionary of the Internet* (Ince 2009), which states that an adhocratic organization is one "that does not rely on job descriptions, hierarchy, standards, and procedures; rather, workers in the company carry out tasks because they need to be done."

Adhocracies encourage members and teams to aggregate and disperse according to need. While individuals do have some roles to play, these roles are not permanent. An adhocratic order is a type of organic structure within which ties of dependence and collaboration exist. Yet, interpersonal connections and the structure itself are flexible. Leaders may, at times, be quite distant from the rest of the group in terms of collaboration, yet this is all temporary and liable to change, as leaders depart for periods of time and new leaders constantly emerge.

In terms of operationalization, adhocracy can be conceptualized in several ways. In this book, adhocracy is considered to be both a broader framework and a specific organizational configuration. As a framework, adhocracy refers to situations in which the elite group's composition is both changing and resilient. In more specific terms, adhocracy is an organizational configuration. As such, it is described as a multidimensional phenomenon, which can be measured in terms of its network or organizational form.

In Part I of this volume, we look at adhocracy through the first lens—that of elite churn—which we may describe in more vernacular terms as the "stickiness" of elite members. We take a dynamic approach, addressing the manner in which changes in stickiness, or increases and decreases in turnover rates among members of the elite group, are associated with changes in structuration (entropy). In other words, we determine whether decreasing entropy is associated with a decrease in elite turnover.

In Part II, we return to the concept of adhocracy, describing it as one of several possible organizational configurations. This portion of the text examines adhocracy through the perspective of entropy (structuration) and four other collaborative factors: collaborative attractiveness, collaborative extroversion, communication flows, and partner choice.

Returning, for the moment, to the first part of the analysis, recall that the inverse of entropy is order. Therefore, the lower the entropy and the greater the structuration, the more likely adhocracy may be to consolidate and become a quasi-stable form of leadership. To think of it in other terms, if the system becomes more stable and predictable because an elite group becomes increasingly dominant over time, then it stands to reason that those highly productive members who would be considered elite are likely to maintain their stable positions over a long duration. Yet, this resilience over time is not absolute. On the contrary, elite stickiness is temporally limited, a conjecture that is addressed at length in the chapters to come.

3.5 Research Questions

Given these premises, our first primary mission is to investigate the evolution of structuration over time and to determine the inflection points in its evolution, in order to pinpoint any discrete phases that are specific to adhocratic communities of practice.

In formal terms, we propose an evolutionary model for detecting the emergence of functional roles and of adhocratic leadership mechanisms. The temporal dependency between elite stickiness and system inequality, which offers evidence of a causal relationship, must first be assessed. We then examine how "sticky" the elites are in distinct periods of organizational development and, as levels of inequality increase, whether members of the elite are indeed more likely to linger in high contributor positions, as suggested above. Finally, we evaluate the direction of this temporal relationship—that is, whether increasing stickiness precedes growing inequality or vice versa.

After connecting stickiness to group structuration as described above, we subsequently examine the extent to which promotion to elite status is conditional on one's collaborative connections to existing members of the elite. Two scenarios are possible. Collaboration with members of the existing elite may contribute toward eventually being counted among their ranks, or ascension to elite status may be the product of one's own efforts regardless of interpersonal connections. The first scenario is more common in everyday, bureaucratic, or otherwise hierarchical organizational configurations, while the second would suggest an adhocratic configuration.

The explanatory mechanism proposed above is encapsulated by the following five research questions, which will offer the first in-depth exploration of the emergence of adhocracy on Wikipedia:

1. Is there a distinct group of highly productive (elite) users on Wikipedia, and if so, to what extent do they dominate contributions to the project over time?
2. What is the social mobility (or its inverse, elite "stickiness") of functional leaders on Wikipedia over time?
3. Are there distinct growth phases determined by the emergence of functional leaders in Wikipedia's first 9 years of existence (2001–2010)?
4. What is the relationship between global inequality, or social structuration, and elite stickiness?
5. Is promotion to the elite group a function of prior interactions with elites?

Before presenting the data analysis that was used to answer the questions above, we would like to expand some more on the topic of entropy and its use as a measure for social structuration, as this is an essential component of our approach that deserves in-depth attention. Therefore, the next chapter is dedicated to this theoretical topic. Readers already familiar with Shannon's work may skip that chapter and proceed directly to the analysis in Chap. 5.

References

Arazy O, Nov O (2010) Determinants of Wikipedia quality: the roles of global and local contribution inequality. In: Inkpen K, Gutwin C, Tang J (eds) Proceedings of the 2010 ACM conference on computer supported cooperative work. ACM Press, New York

Arazy O, Ortega F, Nov O, Yeo L, Balila A (2015) Functional roles and career paths in Wikipedia. In: Cosley D, Forte A, Ciolfi L, McDonald D (eds) Proceedings of the 18th ACM conference on computer supported cooperative work & social computing. ACM Press, New York

Backstrom L, Huttenlocher D, Kleinberg J, Lan X (2006) Group formation in large social net-
    works: membership, growth, and evolution. In: Eliassi-Rad T, Ungar L, Craven M, Gunopulos
    D (eds) Proceedings of the 12th ACM SIGKDD international conference on knowledge discov-
    ery and data mining. ACM Press, New York
Bailey KD (1990) Social entropy theory. State University of New York Press, Albany
Bales RF, Slater PE (1956) Role differentiation in small decision-making groups. In: Parsons T,
    Bales RF (eds) Family, socialization and interaction process. Routledge, London, pp 259–306
Barabási A-L (2014) Linked: how everything is connected to everything else and what it means for
    business, science, and everyday life. Basic Books, New York
Benkler Y, Shaw A, Hill BM (2015) Peer production: a modality of collective intelligence. In:
    Malone TW, Bernstein MS (eds) The collective intelligence handbook. MIT Press, Cambridge
Blau PM (1970) A formal theory of differentiation in organizations. Am Sociol Rev 35:201–218
Blau PM (1977) A macrosociological theory of social structure. Am J Sociol 83:26–54
Bureau of Labor Statistics (2016) American Time Use Survey. http://stats.bls.gov/news.release/
    atus.toc.htm. Accessed 13 Oct 2016
Cummings TG, Worley CG (2014) Organization development and change. Cengage Learning,
    Stamford
Dellarocas C, Sutanto J, Grigore M, Tarigan B (2014) Understanding the "few that matter" in
    online social production communities: The case of Wikipedia. Paper presented at the 2014
    winter conference on business intelligence, Snowbird, 27 Feb-1 Mar 2014
Georgescu-Roegen N (1971) The entropy law and the economic problem. Harvard University
    Press, Cambridge
Halfaker A, Kittur A, Kraut R, Riedl J (2009) A jury of your peers: quality, experience and own-
    ership in Wikipedia. In: Riehle D, Bruckman A (eds) WikiSym 2009: proceedings of the 5th
    annual symposium on wikis and open collaboration. ACM Press, New York
Halfaker A, Geiger RS, Morgan JT, Riedl J (2012) The rise and decline of an open collabora-
    tion system: how Wikipedia's reaction to popularity is causing its decline. Am Behav Sci
    57(5):664–688
Ince D (2009) Adhocracy. In: Ince D (ed) Dictionary of the Internet. Oxford Reference. http://www.
    oxfordreference.com/view/10.1093/acref/9780199571444.001.0001/acref-9780199571444-e-
    50?rskey=v6jNsn&result=61. Accessed 31 Jan 2017
Javanmardi S, McDonald DW, Lopes CV (2011) Vandalism detection in Wikipedia: a high-
    performing, feature-rich model and its reduction through lasso. In: Ortega F, Forte A (eds)
    Proceedings of the 7th international symposium on wikis and open collaboration. ACM Press,
    New York
Jullien N (2012) What we know about Wikipedia. A review of the literature analyzing the
    project(s). Available via SSRN. https://papers.ssrn.com/sol3/Delivery.cfm/SSRN_ID2308346_
    code728676.pdf?abstractid=2053597&mirid=1. Accessed 31 Jan 2017
Kriplean T, Beschastnikh I, McDonald DW (2008) Articulations of wikiwork: uncovering valued
    work in Wikipedia through barnstars. In: Begole B, McDonald DW (eds) Proceedings of the
    2008 ACM conference on computer supported cooperative work. ACM Press, New York
Lewin K (1947) Quasi-stationary social equilibria and the problem of permanent change. In: Burke
    WW, Lake DG, Paine JW (eds) Organization change: a comprehensive reader. Jossey-Bass,
    San Francisco, pp 73–78
Leydesdorff L (2002) Indicators of structural change in the dynamics of science: entropy statistics
    of the SCI journal citation reports. Scientometrics 53(1):131–159
Matei SA, Bertino E, Zhu M, Liu C, Si L, Britt BC (2015) A research agenda for the study of
    entropic social structural evolution, functional roles, adhocratic leadership styles, and credibil-
    ity in online organizations and knowledge markets. In: Bertino E, Matei SA (eds) Roles, trust,
    and reputation in social media knowledge markets: theory and methods. Springer, New York,
    pp 3–33
Mintzberg H (1979) The structuring of organizations: a synthesis of the research. Prentice Hall,
    Englewood Cliffs

Muchnik L, Pei S, Parra LC, Reis SDS, Andrade JS Jr, Havlin S, Makse HA (2013) Origins of power-law degree distribution in the heterogeneity of human activity in social networks. Sci Rep 3:1783. doi:10.1038/srep01783

Newman MEJ (2005) Power laws, Pareto distributions and Zipf's law. Contemp Phys 46(5):323–351

Nielsen J (2006) The 90-9-1 rule for participation inequality in social media and online communities. https://www.nngroup.com/articles/participation-inequality. Accessed 31 Jan 2017

Nov O (2007) What motivates Wikipedians? Commun ACM 50(11):60–64

O'Connor M (1991) Entropy, structure, and organisational change. Ecol Econ 3(2):95–122

Osgood CE, Wilson KV (1961) Some terms and associated measures for talking about human communication. Institute of Communications Research, University of Illinois, Urbana

Panciera K, Halfaker A, Terveen L (2009) Wikipedians are born, not made: a study of power editors on Wikipedia. In: Teasley S, Havn E, Prinz W, Lutters W (eds) Proceedings of the ACM 2009 international conference on supporting group work. ACM Press, New York

Pareto V (1906) Manuale di economia politica. Piccola Biblioteca Scientifica, Milan. English edition: Pareto V (1971) Manual of political economy (trans: Schweir AS). MacMillan: London

Plowman DA, Solansky S, Beck TE, Baker L, Kulkarni M (2007) The role of leadership in emergent, self-organization. Leadership Quart 18(4):341–356

Preece J, Shneiderman B (2009) The reader-to-leader framework: motivating technology-mediated social participation. AIS Trans Hum-Comput Interact 1(1):13–32

Sauermann H, Franzoni C (2015) Crowd science user contribution patterns and their implications. P Natl Acad Sci USA 112(3):679–684

Schramm W (1955) Information theory and mass communication. J Mass Comm 32(2):131–146

Seife C (2007) Decoding the universe: how the new science of information is explaining everything in the cosmos, from our brains to black holes, reprint. Penguin Books, New York

Shafiq MZ, Ilyas MU, Liu AX, Radha H (2013) Identifying leaders and followers in online social networks. IEEE J Sel Area Comm 31(9):618–628

Shannon CE, Weaver W (1948) The mathematical theory of communication. University of Illinois Press, Urbana

Toffler A (1970) Future shock. Random House, New York

Tushman ML, Romanelli E (2008) Organizational evolution: a metamorphosis model of convergence and reorientation. In: Burke WW, Lake DG, Paine JW (eds) Organization change: a comprehensive reader. Jossey-Bass, San Francisco, pp 174–225

Van de Ven AH, Poole MS (1995) Explaining development and change in organizations. Acad Manag Rev 20:510–540

Waterman RH Jr (1993) Adhocracy: the power to change. W. W. Norton, New York

Welser HT, Cosley D, Kossinets G, Lin A, Dokshin F, Gay G, Smith M (2011) Finding social roles in Wikipedia. In: Proceedings of the 2011 iConference, Seattle, 8–11 Feb 2011

Wenger E (1998) Communities of practice: learning, meaning, and identity. Cambridge University Press, Cambridge

Chapter 4
Social Structuration Online: Entropy and Social Systems

4.1 Introduction

One of the most important goals of the present volume is to define and relate group structuration to other online organizational and interactional phenomena. Although structuration is a high-level concept that may hold different meanings for different people, within this research, the concept is quite simple and clear. In brief, structuration is equated with the concept of "signal" in information systems, as defined by Shannon and Weaver (1948). Structuration is meaningful order, so by Shannon's logic, structure is the opposite of entropy. Since structure is measured using entropy, we may say that structure increases as the observed value of entropy decreases. Conceptually, this means that structure is captured in the negative by observing the degree to which the system is not random (noisy or disordered).

Shannon's fundamental insight in the information sciences—a simple formula and metric (entropy) that can be used to measure the presence of order and meaning—is extended in this volume to social phenomena. How do we propose to do this? In the following text, we outline our method in detail.[1] For readers familiar with Shannon's work and his use of entropy to define and measure order in information systems, this chapter might cover known territory. However, its role is not to merely state the obvious, but to advance the conversation from communication and information processing to social phenomena as a whole. In this respect, it is an essential step in our argument.

The use of entropy to detect social structuration in online groups may strike some as counterintuitive, as our approach is premised on the assumption that structuration appears when some elements, behaviors, symbols, or actors in a system are more likely to be present or to interact than others. This contradicts some earlier notions about online environments, which were viewed in an idealistic light and promoted on the basis of doing precisely the opposite: namely, equalizing interaction rates via

---

[1] The present chapter expands and adapts Matei et al. (2010).

© Springer International Publishing AG 2017
S.A. Matei, B.C. Britt, *Structural Differentiation in Social Media*, Lecture Notes in Social Networks, DOI 10.1007/978-3-319-64425-7_4

the uniquely egalitarian capabilities of computer-mediated communication. To put it another way, online interaction has often been seen as a panacea for real-world inequalities, as the social order exhibited in online media was supposed to be more egalitarian, more even, and more interactive (Berman and Weitzer 1997; Braman 1994; Kiesler et al. 1984; Licklider and Taylor 1968; Rheingold 2000; Sclove 1995; Sproull and Kiesler 1991).

This egalitarianism was premised on a number of characteristics considered to be intrinsic to online communication technologies, the most important of which was the functional equality of the nodes, human or technological, which constitute any computer network (Hauben 1996). This assumption naturally led the first generation of Internet and new media researchers to believe that online groups might exhibit egalitarian characteristics (Hiltz 1984; Hiltz and Turoff 1978; Kiesler and Sproull 1992). Many further felt that the assumed and quasi-necessary equality of online social systems would hinder or even prevent majorities from becoming tyrannical (Grossman 1995; Hiltz and Turoff 1978) and that the egalitarian nature of the medium would create new avenues for expressing non-mainstream views (Myers 1987; Turkle 1995).

Within this set of interlocked expectations was something akin to a Holy Grail of collaboration: a method for producing rich and diverse knowledge that grows out of the egalitarian efforts of the many, including the ignored, marginalized, and noninvolved voices. This utopian view was intensely and broadly popularized through books such as *The Wisdom of the Crowds* (Surowiecki 2004), *Out of Control* (Kelly 1995), *The Wealth of Networks* (Benkler 2006), and *Emergence* (Johnson 2001).

However, over the past two decades, it has become increasingly apparent that online social groups, including those defined by the latest wave of social media, are in fact quite skewed, with a few contributors responsible for a sizable majority of contributions. This observation clashes, at least in relative if not absolute terms, with the equality that commentators expected to observe. As such, most reflections on such findings have tended to explain them away rather than to deal with inequality as a widely recurring, unavoidable fact. In general, the early literature simply avoided confronting the issue of inequality on its own territory as a constitutive and generative phenomenon. Although methods for characterizing inequality and diversity in online environments were present in earlier research, they were used only episodically and as "pass-through" mechanisms to address other, unrelated issues; the existence of inequality itself was typically viewed as a nonessential by-product (Kittur et al. 2009).

In brief, until recently (Shaw and Hill 2014), inequality and its measurement were not considered to be central, significant issues warranting in-depth investigation in their own right. By largely neglecting this matter, researchers have unwittingly neglected decades-old studies on uneven distributions in a variety of domains related to communication, from programming and agenda-setting (Chaffee and Wilson 1977; Dominick and Pearce 1976) to some of Osgood and Wilson's (1961) and Schramm's (1955) polymath considerations about entropy as a communicative phenomenon. By neglecting the inequality of distributions in social phenomena, we have lost an important point of purchase, not only in the conversation about equality but also in the more significant discussion about the relationship between equal distributions, randomness, entropy, and structuration. This is particularly important when considering online communication, where we have yet to directly and objectively engage structuration and inequality as related system-level phenomena.

In the present volume, we restart this conversation, illustrating how and why online inequality should be a focal point for further study. Inequality, within certain limits, is more than a side effect or a statistical artifact. Rather, it betrays important constitutive phenomena that reveal the mechanisms through which online groups, and social groups more generally, structure themselves.

4.2 Diversity and Social Entropy: A Neglected Tradition

Outside of the communication sciences, the process of characterizing and quantifying a system by assessing how evenly (or randomly) its elements or states are distributed is a relatively mature theoretical and methodological concern (Seife 2007). In physics, for instance, this probabilistic approach is fundamental for understanding system behavior. In this context, entropy is largely used as a measure of system organization or, in other cases, of uncertainty.

The concept was introduced to the social sciences relatively early by Shannon and Weaver (1948), who proposed "entropy" as a central measure of information systems and, therefore, of information itself. It should be remembered that Shannon was, in fact, the scholar responsible for setting information theory on sound methodological grounds, rigorously defining the construct of information and associating it with the nonrandom states of the symbols or carriers in a communication system. For Shannon, communicative acts that carry meaning were characterized by two factors: redundancy and organization. The lower the level of randomness in a given communication act and, therefore, the higher the degree of order, the more likely that act is to carry meaning. In other words, the act of communication can be characterized as "information" or as a "signal" that is distinguishable from meaningless "noise." This is a crucial insight, since Shannon's conceptualization explicitly equates order with nonrandomness.

As mentioned, Shannon and Weaver borrowed the conceptual and mathematical tools needed to describe information load, and by extension the organization of a communicative system, from physics, where the presence or absence of order in a system is termed its level of "entropy." This construct measures how diverse a system is in terms of its constitutive elements. When the elements are present in an equal proportion, they present a maximum level of diversity, so the system is said to have a high level of entropy. When the elements are imbalanced such that some are more prevalent than others, diversity and therefore entropy are low, and we would say that the system is "ordered."

Intuitively, a gallon of water, in which the molecules float freely, each being able to occupy any location in the container, is less organized (and has higher entropy) than the same gallon of water frozen into an ice cube, in which each molecule is bound to a certain spot in the crystalline structure of the frozen water. Therefore, we may think of ice crystals as organized water, so we would characterize ice as having a lower level of entropy than liquid water. This is both metaphorically and scientifically true. Chemistry and physics have, in fact, articulated this idea as the "entropy of fusion": melting a substance always involves an increase in entropy.

With that said, entropy in social systems is defined not as a physical but as a socio-communicative concept, so it requires a modified interpretation. We start from the same idea: disorganized systems are random, so by the logic of probabilities, their elements are more likely to be in the same states as one another. If these elements are people or symbols, systems that distribute them equally are disorganized. Imagine, for instance, a message in which all possible symbols are equally likely to occur. This message is more likely to be nonsensical because it appears (and typically is) disorganized. A text that uses "a" and "z" with equal frequency is probably garbage since there are very few words that include the letter "z."

When observing social systems, if we detect a situation in which all individuals communicate at the same rate, it probably means that they are communicating to no one in particular or that they are communicating only to themselves. Why is this so? Consider the following. The members of a group start talking about a given topic. However, they cannot all talk at the same time. A true communication act requires at least two partners, one of whom should keep quiet. Thus, when a group's members engage in communication among one another, no more than half of their potential to "send" raw communication can be realized at any given moment. Further, if we take sending potential to be similar to "raw communicative potential," we realize that a true act of communication is maximized socially when this potential is only partially realized individually.

As a consequence, when a group is communicating, the degree to which its members send information—and, therefore, their raw communication rates—cannot be absolutely even at any moment. A snapshot of a true communication process at any given moment will reveal an imbalanced communication process. Some talk, while others listen. If we focus on sending alone, and if we consider a hypothetical situation where communication is dyadic and the sending rates among the members that speak relatively equal, the unevenness of communication would be at least 50% greater than what chance alone would predict. In other words, the entropy of communication should be at least half of the maximum possible value that can be obtained when everybody communicates equally.

The lesson here is that absolute egalitarianism is hardly the ideal state for collaboration. Rather, not only is it unrealistic, but it is essentially impossible if any actual communicative and collaborative work is to occur. Social systems demand a degree of structural order in order to function, so a degree of role differentiation—even if those roles are as simple as "talkers" and "listeners"—is an essential prerequisite and one that will inevitably preclude the egalitarianism that some would naïvely prefer to pursue. Without such differentiation, we are left with nothing more than a cluster of people talking to themselves. If everyone is a "talker," then no one is listening.

All told, a perfectly equal distribution of anything in human or communicative affairs betrays not the presence but the absence of order, which has numerous practical implications. Furthermore, by using a mathematical measure of distribution such as entropy, we can determine at a glance whether or not a system is organized. By extension, taking multiple measurements allows us to directly observe whether the system is growing more or less organized over time.

Yet, scholars of communication, sociology, and other related fields have neglected the potential benefits and implications of employing the entropy concept. In their absence, scientists in other disciplines have more deeply explored the issues of diversity and entropy, developing very sophisticated statistical methods for identifying their magnitude within a given context and for assessing their comparability between systems. Economics, geography, and environmental sciences are just a few of the disciplines that have modified and refined the concept of entropy to their own ends (McDonald and Dimmick 2003), with examples including adaptations of the entropy measure to describe the degree of organization or diversity of firms, species, features, etc. in a given population, geographic area, or society (Maignan et al. 2003).

These methods and the theoretical implications of their conceptualization, both of which will be discussed later in this chapter, are now ripe to be incorporated in the communication discipline, especially among those researchers focusing on the emergence and impact of new online communication environments. Communication technologies present an especially great need to measure and understand entropy and its opposite order. After all, online environments are widely considered to be intrinsically egalitarian, and by a leap of "sociological imagination," one may be tempted to think that online order *is* or *should be* driven by equality. Axiological ideals are thus extended to ontological realities: the value expectation that egalitarianism should prevail in online collaboration and communication extends a moral ideal to an alternative vision for creating social order and social coordination (Johnson 2001; Rheingold 2002).

Online environments are generally believed, especially by several key early observers, to have the ability to self-organize and to create "emergent" orders precisely because all participants are equal. The idea posed is that individual actors within a system are no different than neurons within the brain, so the "hive mind" of users adopts whatever order is necessary at a given moment in time rather than adhering to an arbitrarily imposed hierarchy (Johnson 2001). Here, the presupposition is that egalitarianism is not only a normative ideal but also one demanded by efficiency. The more decentralized and egalitarian the interaction and the more massive the collaboration, the argument goes, the more likely the online systems are to generate solutions that hierarchical, top-down methods of control and coordination could not (Raymond 2001). Yet these claims cannot be verified and the nature of these phenomena cannot be understood until we improve our measurement of the evolution of collaborative organization over time, thereby allowing us to ascertain whether increased egalitarianism is desirable or even, for that matter, feasible.

As collaborative communication environments have become mainstream tools, and as researchers have increasingly relied upon presuppositions of social structure rather than objective data, the low awareness (and, in some circles, the total lack of acknowledgment) of online structuration measures that employ entropy has become more and more glaring, and the need to redress this situation has grown increasingly urgent. Any possible first step toward resolving our lack of concepts and tools to understand structuration must involve some methodological strategies and theoretical principles that can be jointly used to operationalize, measure, and understand entropy in online collaborative and communication environments.

To illustrate the practical implications of our argument, we will discuss the collaborative diversity issues raised by one of the most intriguing technologies that have emerged in the field of online collaboration, Wikipedia. We will illustrate our main points about the utility and operationalization of entropy with a number of examples related to this online project. These examples function not only as a self-contained conversation, but more importantly, they set the stage for our more in-depth study of the project's structuration over time, of its adhocratic organizational strategies, and of the emergent online social phenomena that these processes bring to light.

To illustrate our main points about using entropy in the study of online environments, we start by providing a brief overview of Wikipedia's inner workings, which will serve as our main research laboratory. This will be followed by a discussion about some specific methodological and theoretical avenues related to diversity and entropy. We will conclude with an example of how these principles and methodologies can be used in practice for measuring collaborative diversity and entropy on Wikipedia over time, with the implications derived from this example yielding important insights for future research on online collaborative environments.

4.3 Wikipedia

One of the most significant technologies that have emerged in the last several years is Wikipedia, an online collaborative encyclopedia created outside of traditional authorship, editorial, and copyright constraints, which is built around the very idea of equality and diversity of contributions. As an "open content" repository of encyclopedic knowledge, Wikipedia was designed from its foundations as a collective and distributed effort (Lih 2009; Wikipedia 2016a).

Akin to the open source software movement, described in the famous essay "The Cathedral and the Bazaar" (Raymond 2001), Wikipedia relies, as its name suggests, on the wiki publishing paradigm (Leuf and Cunningham 2001). Wikis[2] are web-based collective and nondirected online publication systems shaped on the model of the Portland Pattern Repository. The first wiki repository, launched by programmer Ward Cunningham in 1995, served as a historical record of computer programming ideas. Like other wiki systems that followed, Ward's wiki used an open web-based

---

[2] Wiki is the Hawaiian word for "quick" (Andrews 1865), implying the speed with which changes can be implemented (Cunningham 2005).

editing interface, which allowed any visitor to the website to add, delete, and publish content at will.

The wiki idea was transferred from computer programming to knowledge production in 2000, as an attempt to improve the editorial process of an online peer-reviewed encyclopedia (Nupedia). Although the peer-reviewed expectations limited the Nupedia project, it laid the groundwork for Wikipedia, which was founded by Jimmy Wales, a businessman with a Master's degree in economics, and by Larry Sanger, an academic philosopher. While Nupedia ceased operation in 2003 after having produced only 25 peer-reviewed articles, Wikipedia has experienced explosive growth: as of 2016, Wikipedia has tens of millions of contributors who have jointly developed over five million articles in English alone, in addition to those in over 100 other languages (Wikipedia 2016c).

Wikipedia's simple web interface allows any visitor to read and, if he or she disagrees or finds the content inaccurate or insufficient, to immediately alter or change its entries (Wikipedia 2017b). The edits are subjected only to limited editorial gatekeeping. For some articles, a Wikipedia member must be registered for a number of days before her or his changes will be immediately accepted by the system. The vast majority of articles, however, may be freely edited, even by unregistered users.

As for the editorial process itself, on each entry page, there is an "edit" button; when pressed, it switches the page from "display" to "edit" mode. The reader can then make any desired changes immediately, and, for most pages, this means that alterations are recorded almost instantaneously. In fact, initially, all Wikipedia articles could initially be edited with no restriction. However, several incidents—most notably that of John Seigenthaler, a former Robert Kennedy aide who was falsely accused on Wikipedia of participating in the conspiracy that killed his former employer (Seigenthaler 2005)—necessitated the invocation of a "protection" policy (Hafner 2006) for some high-profile pages, such as those dedicated to George W. Bush, Jesus, or Adolf Hitler, which are among the most controversial on Wikipedia (Goodfellow 2016; Wikipedia 2017a). This means that only administrators and selected users can edit "protected" articles. Again, however, most articles are editable by any user, even those who have not registered an account on Wikipedia. Those individuals who do register and complete nominal editorial work gain even more leeway in editing the site.

Despite newer policies facilitating article protection, the Wikipedia editorial system remains relatively open. The idea is still to make any reader into a coauthor of the Wikipedia project and, when possible, to give everyone the chance to contribute equally. Although Wikipedia utilizes a code of conduct, which dictates the manner in which content should be changed (Wikipedia 2017b), this set of guidelines is limited in scope and, in many cases, essentially unenforceable. Its main requirement is that each contributor abides by a "neutral point of view" writing policy (Wikipedia 2016b). This means that the contributor should avoid making personal comments or otherwise injecting judgments or opinions about the various perspectives on a specific topic. Disputes regarding the neutrality of a specific article may be settled through arbitration, but this is a protracted process that is generally avoided (Matei and Dobrescu 2011).

More important than the code of conduct is the expectation that even when partisan interests or inaccuracies seep into the read-write mechanism, the system itself will provide the means for quick redress. It is assumed that communal editing, by its open nature, will ensure continuous vigilance and general objectivity. In other words, the rationale for Wikipedia's editing process hinges on the assumption that exposing an article to many users will result in self-correction. In theory, as soon as a biased contribution is posted, a thousand eyes will spot and correct it (Sanger 2001), or to quote Linus' law of computer programming, "Given enough eyeballs, all bugs are shallow" (Raymond 2001).

Yet the editorial mechanism described above relies, to a significant degree, on the expectation that the collaborative efforts will be egalitarian. The basic premise of a wiki system is that knowledge will be more abundant, reliable, and useful when it incorporates highly diverse inputs and viewpoints from a large number of contributors who are all offering significant ideas and contributions, a notion that demands at least a degree of equality among contributors. Further, the egalitarianism incorporated in the method of publication itself is expected to motivate contributors, driving the generation of knowledge that offers quality and reliability that is equivalent, if not superior, to that of knowledge developed from more stratified social systems. Of course, these are normative expectations. The observable facts, as mentioned above and reiterated below, demand a different explanatory mechanism for Wikipedia's staying power.

There are many issues associated with claims that egalitarian systems will self-correct, the most obvious of which is that, after experimenting with radical openness, Wikipedia itself was compelled to switch to a mix of openness and controlled access. Even more important, however, is the fact that equality might not have been present in the first place, despite the deeply ingrained assumptions to the contrary. After all, even though synthetic indicators of equality were used sporadically and only for descriptive purposes, there were nonetheless some early indications that the collaborative process was quite skewed, even at the early stages of the project (Ortega et al. 2008). When one considered the success of Wikipedia—which has often been attributed to its supposed egalitarianism—the possibility that this perceived equality was a mere illusion from the start has significant ramifications for online social structuration and collaboration, both on Wikipedia and elsewhere.

With that in mind, in the following text, we outline our approach to measure equality, including a number of theoretical and statistical considerations that need to be addressed when measuring the equality of participation in online environments in general and on Wikipedia in particular. However, as has already been mentioned, this is not a goal in and by itself. Rather, it is a means toward an end, which is to explore the issue of structuration, to which the rest of the volume will be devoted.

With that in mind, the goal of the rest of this chapter is threefold. First, we aim to dispel the naïve understanding of online social dynamics, particularly as it relates to the frequent claims of implicit egalitarianism. Second, we offer a justification of our application of Shannon's approach to information to serve as a social indicator and a proxy for structuration. Third, we lay the groundwork for a more in-depth analysis of the relationship between general system-level structuration and local elite resilience. Entropy is, as we have already noted and will further explain below, the core indicator for tracking Wikipedia's evolution over time. Thus, the choice of this metric and its significance deserves due attention.

4.4   Determinants of Structuration

Since the measures that we are introducing to address both claims of equality and our own perspective on structuration are relatively new to the analysis of social media, at least in the particular implementations that we propose, let us start by briefly explaining their statistical characteristics, methodological advantages, and potential shortcomings.

The discussion starts with the conceptual idea of the equal representation of elements in a system. When equal representation is present, there is maximum diversity in the variation of qualities or attributes within a given system (in our case, the online collaborative environment). By this definition, diversity is not necessarily a cultural or even a social ideal; rather, it is a mere instrument of measurement. In fact, although systems that are composed of all possible elements in equal proportions would have a maximum level of diversity by this definition, which emphasizes equal representation, they would not be "diverse" in the traditional sense of the word, which implies differences and, oftentimes, the presence or mitigation of minority groups and viewpoints. To put it another way, our measurement of diversity does not include any evaluative meaning. It does not signal, as is the case in common parlance, that there is a "healthy combination" or a "good" variety of characteristics and elements, each represented to maximize their potential and together to enhance the value of the whole. Diversity as a measurement tool in this context is maximized only when the elements are randomly distributed, thus reflecting a low level of structuration or none at all.

In this particular understanding, diversity is a measurement technique used to address other conceptual issues rather than merely an assumed ideal end state. We use the term in a technical sense, similar to that employed by other sciences. For example, in the contexts of ecology and biology this concept is known as biodiversity, which is defined as a variety of life forms and which may be measured within a given ecological community (Maignan et al. 2003). In economics, various diversity measures are used to evaluate the structure of an industry or of a geographically situated industrial or business environment (Stigler 1983). When applied to social contexts, the notion of diversity commonly refers to the presence of a variety of opinions, cultures, ethnic groups, and socioeconomic characteristics (Maignan et al. 2003; McDonald and Dimmick 2003). Due to its wide utilization, the operationalization of the diversity concept tends to vary from discipline to discipline. For example, biodiversity in ecology includes diversity both within species and among species, as well as comparative diversity among ecosystems. The definition of species itself is also dependent on research contexts, which necessitates additional definitions of diversity as it relates to species.

The common thread uniting all of these uses of diversity, however, is that maximum diversity is not a desired end state or even a possible one. For instance, no ecosystem can survive if all species are equally represented. As animal species feed on each other or on plant species, the ones at the top of the food chain need to be less prevalent than the ones that are at the bottom of the chain. An ecosystem in which the number of foxes and rabbits is equal will collapse, as there will be too many foxes for the available number of rabbits. The exhaustion of the rabbit population will be promptly followed by the starvation of the foxes.

The same applies in economic systems. Firms at the end of the production chain are fewer than the ones at the bottom. An economy with an equal number of units at each level of production would be quite unstructured and could be found only in the earliest stages of human and societal evolution.

In an online collaboration system, we consider the issue of content production. Here, as we will see, when every member does as much work as any other, the work is likely to be very fragmented and thus not very structured. In addition, the lack of proper leadership will lead to problems, as higher transaction costs naturally result in either stagnation in the workflow or poor coordination in work objectives.

Now that the conceptual aspects of diversity in online collaborative systems have been addressed, extending the philosophic issues addressed in Chaps. 2 and 3, the remainder of this chapter focuses on a simpler question: How should we effectively employ diversity and entropy to measure structuration? In Sect. 4.5, we explore this process on a general level, touching on both measurement and conceptual issues, including those related to uncertainty.

4.5   Conceptual Underpinning

Suppose that we have an online communication space (O), which has n number of opinions and m number of members:

$$O = \{O_1, O_2, \ldots O_n\}. \tag{4.1}$$

Let C be a classification of O. The opinions posted on the communication space can be classified by a certain criterion variable. For example, C might be a classification set comprising each participant through number m. Thus,

$$C = \{C_1, C_2, \ldots C_m\}. \tag{4.2}$$

Assume $C_i \cap C_j = 0$ meaning that each opinion in O belongs to only one participant. Thus,

$$\bigcup_{j=1}^{m} C_j = C = O. \tag{4.3}$$

S_i is the share (mathematical proportion) represented by the opinions of the ith individual in the opinion space O:

$$S_i = \frac{|C_i|}{\sum_{j=1}^{m}|C_j|}, \quad \sum_{i=1}^{m}S_i = 1. \tag{4.4}$$

The question that we want to answer, then, is how we can measure and quantify the diversity of contributions, participation, involvement, or presence in this context. As a starting point, suppose that in the online communication space O, there is only one opinion posted by a single participant (Tom).

$$O = \{*_1\} \quad \text{and} \quad P = \{\text{Tom}\} \tag{4.5}$$

where P represents the set of participants in O.

In this scenario, there is no uncertainty about who posted this opinion. It is completely certain that the sole contributor Tom must have been the person who posted his opinion in this communication space—in the absence of other participants, there are no other possibilities—so the online communication space therefore has no diversity.

Now suppose that there is another participant (say, Sara) in O. In other words,

$$O = \{*_1\} \quad \text{and} \quad P = \{\text{Tom, Sara}\}. \tag{4.6}$$

In this situation, the contributions to the online site could be made either by Tom or Sara. Therefore, we necessarily have a degree of uncertainty about contributions. From the perspective of information theory (Shannon and Weaver 1948), it is said that a question like this, which has two possible answers (Tom or Sara), carries 1 bit of information. More generally, if we had m participants in a communication space, the question of who made a given post would have m possible outcomes and would thus carry $\log_2 m$ bits of information (Cover and Thomas 2006). To simplify the matter, this formula yields a rather trivial fact: as more people participate in an online community, social diversity tends to increase and so does uncertainty.

The value of measuring diversity with a mathematical formula becomes clearer when there are many members and many opinions in a communication space. For example:

$$O = \{*_1, *_2, *_3, \#_4\} \quad \text{and} \quad P = \{\text{Tom, Sara}\}.$$
$$C_{\text{Tom}} = \{*_1, *_2, *_3\}, \quad C_{\text{Sara}} = \{\#_4\}. \tag{4.7}$$

In this example (4.7), the star (*) and sharp (#) notations represent opinions. Tom posted three opinions, denoted with stars (*), and Sara posted one, given by a sharp

(#). Tom's posts comprise 75% of the total opinions, with Sara's accounting for the remaining 25%.

To extend the example, we can consider more diverse communication spaces with equal contributions by all members, such as the following:

$$O = \{\Delta_1, \Phi_2, \Omega_3, \Psi_4\}.$$
$$P = \{\text{Tom}, \text{Sara}, \text{Kati}, \text{John}\}. \qquad (4.8)$$
$$C_{\text{Tom}} = \{\Delta_1\}, \; C_{\text{Sara}} = \{\Phi_2\}, \; C_{\text{Kati}} = \{\Omega_3\}, \; C_{\text{John}} = \{\Psi_4\}.$$

The equal number of contributions made by these participants, $\frac{1}{n} = 25\%$, clearly implies that the level of diversity in the communication space given in this new scenario (4.8) is higher than that of the communication space in the previous example (4.7). In general, when we observe a uniform distribution of contributions among all participants in a communication space, then that space has the highest possible level of diversity.

4.5.1   Social Entropy as a Measure of Diversity and Equality

In sum, the diversity of opinions in communication spaces is a function of the number of participants (m) and the shares of participants (S_i). The presence of more participants facilitates more diverse (and equal) participation—and as the contributions made by members of a community become more uniformly distributed, participation necessarily becomes more diverse. In this respect, greater diversity means greater uncertainty and "disorder," which can be conceptually explained as a higher level of "entropy."

So, how can we translate this into a synthetic indicator? We can do it, as Shannon and Weaver suggested, by measuring the relative degree of disorganization found in any system. We may think of disorganization as the random mixing of various elements, whose relative presence should thus be equal. In social systems, disorganization also implies a higher level of uncertainty and unpredictability of contributions. In this "random mixing" scenario, we can also say that the diversity of the system is at a maximum, since all elements are equally (randomly) present and their contributions entirely unpredictable.

With this in mind, we may turn to Shannon's entropy index, which takes a value of 0 when there is absolute order in a given system (one element is prevalent at the expense of all others; contribution is perfectly uncertain) and holds a maximum value—which varies from system to system—when there is perfect disorder and diversity (all elements are equally present, and the source of a given contribution is maximally uncertain). In short, entropy is a synthetic measure that indicates, at a glance, the extent to which the different components of a social or communicative space are well represented.

Mathematically, the entropy of a random variable X (in our case, the proportion of contributions made by different members of a social system) is defined as follows:

$$H(X) = -K \sum_{i=1}^{m} p_i \log_2 p_i, \tag{4.9}$$

where K is a constant.

The value of entropy varies from zero (no entropy) to $\log_2 m$ (maximum entropy).[3]

How do we apply this measure to online collaboration environments? Consider an online communication space in which there is a uniform distribution of contributions by four members: $\left(\frac{1}{4}, \frac{1}{4}, \frac{1}{4}, \frac{1}{4}\right)$. The entropy of this communication space is

$$H(X) = -\sum_{4}^{i=1} S_i \log_2 S_i =$$
$$-\sum_{4}^{i=1} \frac{1}{4} \log_2 \frac{1}{4} = \log_2 4 = 2. \tag{4.10}$$

Now, consider another communication space with four members. Assume that the shares of contributions by these members are unequally distributed: $\left(\frac{1}{2}, \frac{3}{10}, \frac{1}{10}, \frac{1}{10}\right)$. The entropy of this communication space is therefore

$$H(X) = -\sum_{4}^{i=1} S_i \log_2 S_i$$
$$= -\frac{1}{2} \log_2 \frac{1}{2} - \frac{3}{10} \log_2 \frac{3}{10}$$
$$-2\left(\frac{1}{10} \log_2 \frac{1}{10}\right) = 1.69. \tag{4.11}$$

The entropy of the communication space with a uniform distribution of contributions is higher than the one with unequally distributed contributions. Likewise— and this is crucial for the substantive interpretations that we can make with this type of measure—when entropy holds its maximum possible value, uncertainty is also maximized, especially with respect to the potential future states of the system.

---

[3] The present example employs logarithms taken at base 2, which allows entropy to be measured in bits. Other values such as e or 10 could be used for the base instead, as this choice is relatively arbitrary and does not affect the calculation of normalized entropy values or any proportional comparisons between contributors and contributions (Lemay 1999).

4.5.2 Normalized Social Entropy as a Diversity/Equality Measure

Before further developing this conceptual point about uncertainty, we need to make a mathematical observation. Although entropy is an elegant modality to measure diversity in a system, it carries some potential limitations. In particular, entropy, as laid out above, reflects not just one but two system dimensions: richness and evenness (equality). When we collapse them into one index score, there is a loss of information. Moreover, the two dimensions may act in opposition to one another such that very different online communication groups in terms of composition and contributions exhibit similar entropy scores (Balch 2000). It goes without saying that this can lead to some confusion.

As an example, communication space (C_1) has two opinions expressed by two participants. The two participants have contributed an equal share of opinions: $\left(\frac{1}{2},\frac{1}{2}\right)$. In contrast, the second communication space (C_2) has 64 opinions and 8 participants, and the opinions are unequally distributed among the 8 participants: $\left(\frac{1}{2},\frac{1}{4},\frac{1}{8},\frac{1}{16},\frac{1}{64},\frac{1}{64},\frac{1}{64},\frac{1}{64}\right)$. However, the calculation of entropy provides a counterintuitive result: while the entropy of the first communication space (C_1) is 1, the entropy of the second communication space (C_2) is 2. Thus, (C_2) has a higher entropy level despite the fact that its opinions are less evenly distributed compared to (C_1). This is because (C_2) has more participants, so the fact that its contributions are unequally distributed is hidden—the change in richness masks differences in evenness between the two communication spaces. The direct consequence is that the absolute increase in the value of entropy does not reliably indicate how much more (or less) organized a system is compared to another if the number of elements (participants) differs between the two systems.

We may solve this problem by normalizing the entropy values. Doing so enables us to compare the evenness of multiple communication spaces or to assess changes in evenness within one communication space over time, by controlling for the number of elements that composes the system. We achieve this normalization by dividing the absolute entropy score by its maximum, $\log_2 m$, which limits the range of possible entropy values from 0 to 1.

$$H_o = \frac{H}{H_{max}}, \quad 0 \le H_o \le 1, \text{ where}$$

$$H_{max} = \log_2 m.$$

(4.12)

As a side note, normalized entropy is particularly useful for handling the "lurker" problem of studying diversity in online environments. Lurkers are users who do not directly contribute to a communication environment; he or she is just an observer. Lurkers can potentially make the communication environment richer, and their presence may also impact diversity. The challenge, then, is capturing both of these aspects of lurker behavior.

Suppose that there are two communication spaces. In both, the two members who contribute offer equal contributions:

$$O_1 = \{\Delta, \Omega\}, P_1 = \{\text{Tom, Jane}\} \text{ with the share distribution} \left(\frac{1}{2}, \frac{1}{2}\right).$$

$$O_2 = \{\Delta, \Omega\}, P_2 = \{\text{Tom, Jane, Sara}\} \text{ with the share distribution} \left(\frac{1}{2}, \frac{1}{2}, 0\right). \tag{4.13}$$

In addition to Tom and Jane, who contributed equally, the second communication space also includes lurker Sara, who did not contribute to the interaction. Despite this important difference, the non-normalized entropy of the two communication environments is the same:

$$H\left(X_{C_1}\right) = -\sum_3^{i=1} S_i \log_2 S_i = -\frac{1}{2}\log_2\frac{1}{2} - \frac{1}{2}\log_2\frac{1}{2} = 1.$$

$$H\left(X_{C_2}\right) = -\sum_3^{i=1} S_i \log_2 S_i = -\frac{1}{2}\log_2\frac{1}{2} - \frac{1}{2}\log_2\frac{1}{2} - 0\log_2 0 = 1. \tag{4.14}$$

Normalizing the entropy values highlights the presence of the lurker in one of the spaces. For example, the maximum $\log_2 m$ entropy value for C_1 is

$$H_{\max}\left(X_{C_1}\right) = \log_2 2 = 1 \tag{4.15}$$

while its normalized value is

$$H_o\left(X_{C_1}\right) = \frac{H}{H_{\max}} = \frac{1}{1} = 1. \tag{4.16}$$

For C_2, the maximum entropy value is

$$H_{\max}\left(X_{C_2}\right) = \log_2 3 \cong 1.58, \tag{4.17}$$

and, thus, the normalized social entropy of C_2 is

$$H_o\left(X_{C_2}\right) = \frac{H}{H_{\max}} = \frac{1}{1.58} \cong 0.63. \tag{4.18}$$

By comparing the normalized entropy of the two communication spaces, we can see that the first communication space is more diverse—that is, more randomly distributed and less structured—than the second one, because the normalized entropy formula properly accounts for the presence of the lurker.

4.6   Using Entropy to Study Online Collaborative Systems such as Wikipedia

Social entropy can be used to measure changes over time in terms of contribution diversity and, thus, structuration of contributions to Wikipedia articles (entries). When tracked over time, entropy can also be used to measure system-level changes in structuration, beyond individual articles in isolation. We can therefore use values of social entropy to answer two key questions: "How equally distributed are the contributions to Wikipedia among users?" and "How structured are collaborative groups on Wikipedia?"

It should further be noted that our earlier examples assumed that all contributions were equal, but this may not necessarily be the case. After all, in a collaborative system like Wikipedia, some contributions—in this context, revisions to an article—may have much more of an impact than others. (Should we really deem contributing a paragraph of original content to have the same value as correcting a single typographical error?) Such differences need to be addressed in order to properly assess the social structuration of collaborative processes.

With that in mind, the following example illustrates how entropy can be used to determine the evolution of collaborative structuration in specific Wikipedia articles and to do so while accounting for the relative significance of each contribution.

We selected the first few edits made to the Indian film *Naina* (Wikipedia 2008). On May 16, 2005, Hemanshu created the article using 16 words[4]: "*Naina* is a Hindi movie to be released in India in 2005. It stars Urmila Matondkar." On July 4, 2005, an anonymous user (A1)[5] added 15 words at the end of Hemanshu's original entry: "It's genre is horror. It is having great similarities over the English film The Eye." On December 24, 2005, a second anonymous user (A2) contributed 24 more words to the article, and on December 30, a third anonymous user (A3) proofread and edited the article, deleting 21 words and adding 10 of his or her own (in bold).

> *Naina* is a Hindi movie released in India in 2005. It stars Urmila Matondkar. It's genre is horror. It **has many** similarities **with** the English film The Eye. **It's release created controversy** in India because **of** the fact that the lady had eye **transplant** before **experiencing extra sensory perceptions in the film, and that** discouraged many people from **receiving** eye **transplants**.

Finally, in January and February 2006, NilsB and DomLachosicz added external hyperlinks[6] and jointly deleted 13 words and added 12 new words.

---

[4] As an alternative to the number of words added, deleted, or modified, one could instead choose to use the number of characters as a measure of the amount of information contributed. The choice of measures will be addressed at greater length in Chap. 5.

[5] On Wikipedia, registered users have their own screen name that appears in the history page for every article. Peripheral users with no membership have no screen name. However, we may still determine who wrote what using anonymous users' IP addresses, which the wiki makes visible.

[6] Measuring visible text is a relatively straightforward task. In contrast, it is challenging to determine how to count and measure the effort and/or impact of visual content and interactive media such as diagrams, photographs, videos, hyperlinks, and so forth. This conceptual and technical question calls for further scholarly attention and discussion; it is not addressed at length in this volume. The present study considers only visible text in calculations of effort and entropy, as well as various other measures to be discussed later in this volume.

Fig. 4.1 Relative contribution trend

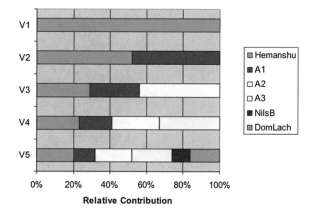

Table 4.1 Social entropy of the Wikipedia article on *Naina*

	Version 1	Version 2	Version 3	Version 4	Version 5
# words (%)[a]	16 (100%)	31 (100%)	55 (100%)	61 (100%)	69 (100%)
Hemanshu	16 (100%)	16 (52%)	16 (29%)	14 (23%)	14 (20%)
A1[b]	–	15 (48%)	15 (27%)	11 (18%)	8 (12%)
A2[b]	–	–	24 (44%)	16 (26%)	14 (20%)
A3[b]	–	–	–	20 (33%)	15 (22%)
NilsB	–	–	–	–	7 (10%)
DomLach	–	–	–	–	11 (16%)
Social entropy[c]	0	0.99	1.55	1.96	2.53
# participants	1	2	3	4	6
Max. entropy[d]	0	1.00	1.58	2.00	2.58
Normalized H[e]	–	0.998	0.980	0.982	0.986

[a]The percentile represents relative shares (S_i) of each contributor in textual content
[b]A1, A2, and A3 each represent an anonymous contributor with no screen name
[c]Social entropy (H) is calculated using the following formula: $H = -\sum_{m}^{i=1} S_i \log_2 S_i$
[d]Maximum entropy (H_{max}) is calculated using the following formula: $H_{max} = \log_2 m$
[e]Normalized social entropy (H_o) is calculated using the following formula: $H_o = H/H_{max}$

Figure 4.1 shows the relative contributions made by the editors who made these initial substantive revisions, with the proportions based on the number of words added or deleted from the text to showcase the impact of a given revision upon the article's content. The changes in these proportions are given over time, with five especially meaningful iterations of the article provided for illustrative purposes.

Likewise, Table 4.1 provides the descriptive and entropy statistics associated with these five successive versions. As shown in Table 4.1, the length of the article substantially increased over time, reaching 69 words by the end of this set of revisions.

As a consequence of the increase in textual contributions and the number of contributors, the social entropy of the article also increased over time. As is evident

Fig. 4.2 Social entropy
trend

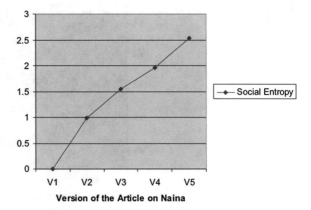

Version of the Article on Naina

Fig. 4.3 Normalized
social entropy trend

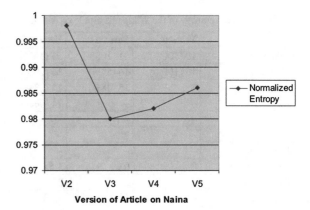

Version of Article on Naina

from Fig. 4.2, social entropy increased from 0 (no diversity) upon the article's creation, when only one user had contributed, to 2.53 in the last version shown, which demonstrates the increasing tendency toward diversity in the collaborative process.

Does the observed increase in diversity, however, represent an increase in randomness and an indication that the editorial process became more uncertain and less structured over time? Or, is it merely a by-product of the growing richness of the communication environment surrounding this article as additional editors joined the collaborative process? To answer this question, we must separate collaborative structuration from mere richness. This can be achieved by converting the raw entropy values using the normalized entropy formula that was previously discussed.

Figure 4.3 shows that normalized entropy did not steadily decline but instead fluctuated over time. The second version of the article[7] had a normalized entropy value of $H_o = 0.998$, which is very close to the maximum value of 1. This means that Hemanshu and the first anonymous user (A1) contributed to almost exactly the

---

[7] Note that the first version was skipped, as normalized entropy cannot be calculated when only one user has contributed since the maximum possible entropy value would be $H_{max} = \log_2 1 = 0$.

same degree. In the third version of the article, the normalized entropy slightly dropped as the second anonymous user (A2) added a relatively large amount of textual information (44% of the words in the article) compared to the other two contributors. Then, in the fourth and fifth versions of the article, the normalized entropy increased as other new contributors (A3, NilsB, and DomLach) participated relatively evenly in this collaborative writing process. Overall, though, the amount of normalized entropy and structuration declined from its initial value, reaching a submaximal level.

Of course, this example has little meaning by itself, since it does not track a sufficient number of edits for us to draw generalizable conclusions about social structuration across collaborative interactions. What this simple example does provide, however, is "proof of concept"—a demonstration of the synthetic characterization of social entropy's evolution over time, through which we may measure the diversity within an article. Further, by comparing the two versions of entropy (raw and normalized), it becomes clear that while the first measure yields only a relatively steady (and uninteresting) increase in entropy over time, the second is capable of capturing subtle fluctuations and variations at a level lower than the maximum possible value.

Moreover, if we compare Fig. 4.2 with Fig. 4.3, we notice that the slope for the increase in raw entropy is far more abrupt than the one for normalized entropy, even after the third version of the article. Observations like this one, along with the fluctuations observed in Fig. 4.3, can facilitate the formulation of tentative hypotheses.

We can speculate, for instance, that the effect of newly added information on diversity will diminish in size as the total amount of textual information increases. In other words, as more and more contributions are made to an article, each successive contribution will tend to have less of an effect on entropy. Consequently, normalized social entropy may tend to level off as the article evolves, eventually reaching a somewhat stable value that represents not merely a meaningless statistical artifact, but a form of structural stabilization within the collaborative process. Such conjectures could be addressed beyond the hypothetical level by directly observing the evolution of social entropy (and collaborative evenness) across the entire collaborative Wikipedia space over a long period of time.

This is, in fact, the approach that we will present in the next few chapters. We used the normalized entropy index to measure the evolution of collaborative structuration across Wikipedia between its inception (2001) and 2010, encompassing nearly a decade of collaborative activity. Contributions to Wikipedia were operationalized as the "amount of effort" or "contribution effort" performed for each unique editorial intervention made to Wikipedia throughout the nine-and-a-half-year period. "Contribution effort" was quantified by Luca de Alfaro and his colleagues at the University of California, Santa Cruz (see de Alfaro 2009 for details) using a "delta" metric, which is used to assess how "far" each successive edit moves a given article from its preceding version. This measure encompasses three factors: how much content was contributed through a particular edit, how much old content was deleted in the process, and how much the existing text was reordered. The specific formula used is

$$d(u,v) = \max(I, D) - 0.5\min(I, D) + M, \qquad (4.19)$$

where

I = total inserted text,
D = total deleted text,
M = total relative change of position (a measure of text reordering).

In effect, the edit "distance" formula (de Alfaro et al. 2010) ranges from zero to infinity, representing the amount of content contributed at each editing iteration and, more importantly, the effort that went into generating said content. Minor revisions, like the correction of typographical errors or simple cut-and-paste operations, count for less than content that is new, changes the flow of the text, or is added at multiple locations in the text. Appendix A provides an in-depth presentation of the methods used for pre-processing the data analyzed in this book.

4.7   What Have We Learned So Far?

Social entropy is a very promising procedural approach for assessing group dynamics, especially in terms of measuring the structuration of contributions in online communication. This is relevant not only for wiki articles but also in other types of virtual communities: social media, newsgroups, email lists, blogs, and so forth. Given this cross domain applicability, a significant impact of this measure would be to rekindle interest in system-level processes and their measurement in online environments.

Social entropy enables us to characterize the status of a system with respect to its level of diversity and structuration in a very parsimonious manner. A wide range of questions—does this communication system develop a structure over time? at what optimal level does the structuration process stop? and so on—become addressable and measurable. Of course, the assumption that open, networked systems would automatically engender egalitarian environments has been relatively naïve. We already suspect that as online communication systems matured, they often became dominated by the few at the expense of the many. With that said, using the approach proposed above, we can transcend mere conjecture and form decisive conclusions about social inequalities. In short, we now have a way to systematically examine these inequalities as more than anecdotal aberrations.

Thus, the need to examine diversity, structuration, and the equality of contributions (or the lack thereof) is also an invitation to use social entropy, not as an end in itself, but as a means toward better understanding the vast array of social and communicative systems in our world. In particular, although it may initially seem intuitive, the claim that open communication systems *should be* egalitarian or tend toward diversity and equality—that is, that their levels of entropy should increase over time—does not make much theoretical sense. Such suppositions contradict what we know about human organization, which is that at least for some periods of growth it becomes more structured and predictable, not less.

Future research should take into account the fact that diversity and equality, although seemingly admirable societal and moral goals, might not always be present or even functionally desirable in the structure of a collaborative group, especially when diversity increases to such an extent that it affects the group's ability to structure itself. Processes related to group structure and functional differentiation of the members, creation and maintenance of group communicative standards, preservation of a given meaning over time, and prevention of ambiguity might be critical forces that promote the group's evolutionary life cycle and growth. Additionally, as groups coalesce and functional hierarchies are formed, entropy may not be maximized but instead reduced or maintained at an optimal level. From this perspective, one of the most important theoretical goals of the social entropy perspective would be to discern what entropy levels are optimal for any given online communicative and collaborative system and, further, to pinpoint which entropy level corresponds to what degree of functional differentiation and hierarchical organization.

Ultimately, what makes the social entropy index so enticing is that it gives researchers a gauge to directly measure and assess levels of collaboration and their significance. Yet, the theoretical and practical ramifications of such a measure extend even further than that. In the chapters to come, we will investigate some of these wide-ranging ramifications as they emerge on Wikipedia.

References

de Alfaro L (2009) Text evolution and author contributions. https://goo.gl/IM7tQG. Accessed 28 Jan 2017

de Alfaro L, Adler B, Pye I (2010) Computing wiki statistics (WikiTrust). http://wikitrust.soe. ucsc.edu/computing-wiki-statistics. Accessed 10 Sept 2012

Andrews L (1865) A dictionary of the Hawaiian language, to which is appended an English-Hawaiian vocabulary and a chronological table of remarkable events. H. M. Whitney, Honolulu

Balch T (2000) Hierarchical social entropy: an information theoretic measure of robot group diversity. Auton Robot 8:209–237

Benkler Y (2006) The wealth of networks: how social production transforms markets and freedom. Yale University Press, New Haven

Berman J, Weitzer DJ (1997) Technology and democracy. Soc Res 64(3):1313–1319

Braman S (1994) The autopoietic state: communication and democratic potential in the net. J Am Soc Inform Sci 45(6):358–368

Chaffee S, Wilson DG (1977) Media-rich, media poor. Two studies of diversity in agenda-holding. Journal Q 54:466–476

Cover TM, Thomas JA (2006) Elements of information theory. Wiley, New York

Cunningham W (2005) Correspondence on the etymology of wiki. http://c2.com/doc/etymology. html. Accessed 28 Jan 2017

Dominick JR, Pearce MC (1976) Trends in network prime-time programming, 1953–1974. J Commun 20:70–80

Goodfellow M (2016) The 6 most controversial edited Wikipedia pages. Independent. http://www. independent.co.uk/life-style/gadgets-and-tech/the-6-most-controversial-edited-wikipedia-pages-a6814756.html. Accessed 31 Jan 2017

Grossman LK (1995) The electronic republic: reshaping democracy in the information age. Viking, New York

Hafner K (2006) Growing Wikipedia refines its "anyone can edit" policy. The New York Times. http://www.nytimes.com/2006/06/17/technology/17wiki.html. Accessed 31 Jan 2017

Hauben M (1996) Netizens: on the history and the impact of the net. http://www.columbia.edu/~hauben/netbook. Accessed 31 Jan 2017

Hiltz SR (1984) Online communities: a case study of the office of the future. Ablex, Norwood

Hiltz SR, Turoff M (1978) The network nation: human communication via computer. Addison-Wesley, Reading

Johnson S (2001) Emergence: the connected lives of ants, brains, cities, and software. Scribner, New York

Kelly K (1995) Out of control: the new biology of machines, social systems and the economic world. Addison-Wesley, Reading

Kiesler S, Sproull L (1992) Group decision making and communication technology. Organ Behav Hum Dec 52:96–123

Kiesler S, Siegel J, McGuire T (1984) Social psychological aspects of computer mediated communication. Am Psychol 39(10):1123–1134

Kittur A, Lee B, Kraut RE (2009) Coordination in collective intelligence: the role of team structure and task interdependence. In: Olsen DR Jr, Arthur RB, Hinckley K, Morris MR, Hudson S, Greenberg S (eds) Proceedings of the SIGCHI conference on human factors in computing systems (CHI 2009). ACM Press, New York

Lemay P (1999) The statistical analysis of dynamics and complexity in psychology: a configurational approach. Dissertation, University of Lausanne

Leuf B, Cunningham W (2001) The Wiki way: quick collaboration on the web. Addison-Wesley, Boston

Licklider JCR, Taylor RW (1968) The computer as a communication device. Sci Technol: Tech Men Manag 76:21–31

Lih A (2009) The Wikipedia revolution: how a bunch of nobodies created the world's greatest encyclopedia. Hyperion, New York

Maignan C, Ottaviano G, Pinelli D, Rullani F (2003) Bio-ecological diversity vs. socio-economic diversity: a comparison of existing measures. Available via SSRN. https://papers.ssrn.com/sol3/Delivery.cfm/SSRN_ID389043_code030325590.pdf?abstractid=389043&mirid=1. Accessed 31 Jan 2017

Matei SA, Dobrescu C (2011) Wikipedia's "neutral point of view": settling conflict through ambiguity. Inform Soc 27(1):40–51

Matei SA, Oh K, Bruno R (2010) Collaboration and communication in online environments: a social entropy approach. In: Oancea M (ed) Comunicare Đi comportament organizational (Communication and organizational behavior). Printech, Bucharest, pp 82–98

McDonald DG, Dimmick J (2003) The conceptualization and measurement of diversity. Commun Res 30:60–79

Myers D (1987) Anonymity is part of the magic: individual manipulation of computer-mediated communication contexts. Qual Sociol 19(3):251–266

Ortega F, Gonzalez-Barahona JM, Robles G (2008) On the inequality of contributions to Wikipedia. In: Proceedings of the 41st Hawaii international conference on system sciences (HICSS '08), IEEE, Washington, DC, 7–10 Jan 2008

Osgood CE, Wilson KV (1961) Some terms and associated measures for talking about human communication. Urbana, Institute of Communications Research, University of Illinois

Raymond ES (2001) The cathedral and the bazaar: musings on Linux and open source by an accidental revolutionary, Revised. O'Reilly, Cambridge

Rheingold H (2000) The virtual community: homesteading on the electronic frontier. MIT Press, Cambridge

Rheingold H (2002) Smart mobs. Perseus Publishing, Cambridge

Sanger L (2001) Wikipedia is wide open. Why is it growing so fast? Why isn't it full of nonsense? Kuro5hin. http://www.kuro5hin.org/story/2001/9/24/43858/2479. Accessed 1 Nov 2008

Schramm W (1955) Information theory and mass communication. J Mass Comm 32(2):131–146

Sclove R (1995) Democracy and technology. Guilford Press, New York

Seife C (2007) Decoding the universe: how the new science of information is explaining everything in the cosmos, from our brains to black holes, Reprint. Penguin Books, New York

Seigenthaler J (2005) A false Wikipedia 'biography.' USA Today. http://www.usatoday.com/news/opinion/editorials/2005-11-29-wikipedia-edit_x.htm. Accessed 31 Jan 2017

Shannon CE, Weaver W (1948) The mathematical theory of communication. University of Illinois Press, Urbana

Shaw A, Hill BM (2014) Laboratories of oligarchy? How the Iron Law extends to peer production. J Commun 64(2):215–238

Sproull L, Kiesler SB (1991) Connections: new ways of working in the networked organization. MIT Press, Cambridge

Stigler GJ (1983) The organization of industry. Richard D. Irwin, Homewood

Surowiecki J (2004) The wisdom of crowds: why the many are smarter than the few and how collective wisdom shapes business, economies, societies, and nations, 1st edn. Doubleday, New York

Turkle S (1995) Life on the screen: identity in the age of the Internet. Simon & Schuster, New York

Wikipedia (2008) Naina. Wikipedia. http://en.wikipedia.org/wiki/Naina. Accessed 30 Oct 2008

Wikipedia (2016a) Wikipedia. Wikipedia. http://en.wikipedia.org/w/index.php?title=Wikipedia. Accessed 28 Oct 2016

Wikipedia (2016b) Wikipedia:Neutral point of view. Wikipedia. http://en.wikipedia.org/w/index.php?title=Wikipedia:Neutral_point_of_view. Accessed 28 Oct 2016

Wikipedia (2016c) Wikipedia:Statistics. Wikipedia. http://en.wikipedia.org/w/index.php?title=Wikipedia:Statistics. Accessed 28 Oct 2016

Wikipedia (2017a) Wikipedia:Most frequently edited pages. Wikipedia. https://en.wikipedia.org/wiki/Wikipedia:Most_frequently_edited_pages. Accessed 2 Jan 2017

Wikipedia (2017b) Wikipedia:Policies and guidelines. Wikipedia. https://en.wikipedia.org/wiki/Wikipedia:Policies_and_guidelines. Accessed 29 Jan 2017

Chapter 5
Analytic Investigation of a Structural Differentiation Model for Social Media Production Groups

5.1 Introduction

In the previous chapters, we articulated the role of evolutionary processes on Wikipedia and similar social media production sites and then outlined the main reasoning behind using entropy as a measure of structuration. The main claim is that, just like many other self-organizing voluntary organizations, online social groups need to overcome the increasing costs of communication and coordination as they continue to grow and evolve, which they do via the emergence of leading contributors who work substantially more than their peers.[1]

By taking on central roles in the production process, emergent leaders provide several services to the group at once. First, they reduce the collaborative paths between the core participants. The contributors who are responsible for most of the contributions only need to keep an eye on each other to learn what is happening around them.

More importantly, their constant presence in a variety of collaborative contexts within the group allows them to act as go-between nodes in the structure of collaboration. This need not be explicit or even deliberate, as a leader's mere presence and the social cues sent through one's contributions signal to the rest of the collaborative team that a central member of the project is about to bring the current collaborative task to closure.

Finally, emergent leaders play a crucial role in shaping the content itself, ensuring continuity, direction, and narrative coherence. Although these elements may, at times, be tainted by partisanship and turf wars, which are typically won by the participants with the most time available to "fight it out," this is an unavoidable cost that must be accepted for the sake of the project's success (Kriplean et al. 2007, 2008).

---

[1] In what follows, we summarily recapitulate the arguments made in the previous chapters. For the sake of readability, we have limited the citations to the minimum necessary. All the claims made in this chapter, however, are richly and completely referenced in Chaps. 2, 3, and 4 of Part I.

© Springer International Publishing AG 2017   69
S.A. Matei, B.C. Britt, *Structural Differentiation in Social Media*, Lecture Notes in Social Networks, DOI 10.1007/978-3-319-64425-7_5

Throughout their interventions, activities, and implicit and explicit communication processes, the leaders lead first and foremost by doing more than their peers, not by merely claiming to be leaders. Their leadership is, as previously mentioned, functional. It is leadership that comes from working harder than everybody else. If the collaborative process on Wikipedia were likened to a marathon, the leaders would be those who could run the fastest over the longest periods of time. Wikipedia and many other social media projects are, after all, endurance contests. As soon as a leader ceases to contribute at a high rate and be present in as many aspects of the project as possible, he or she starts sliding down the metaphorical ladder. This process is captured within the idea of adhocracy, as leading by doing is necessarily characterized by a certain turnover rate. This turnover rate can neither be too low (which would imply leaders who are entrenched regardless of effort) nor too high (suggesting a lack of any consistency whatsoever in the leadership cadre).

In this chapter, we will look at the contours of leadership groups and at their temporal persistence, or stickiness, the latter of which we will assess via their turnover. We will further analyze the relationship between the evolution of group structuration (entropy) and elite stickiness in order to elucidate the role of elites in relation to the larger organization. Finally, we will examine the factors that explain how individuals advance to elite status, thereby adding a microlevel perspective to the meso- and macro-level dynamics under study.

5.2 Methods

5.2.1 Data

The present study investigates Wikipedia's first 9.5 years of editorial activity, spanning the beginning of 2001 through June 2010. Our data includes every single editorial change made to any content article. Revisions to talk or policy pages were not included in the data set, as we chose to focus on the most visible outputs that would have a direct effect on Wikipedia visitors, most of whom are unlikely to view any pages beyond the articles themselves. Data was obtained from the MediaWiki project at https://dumps.wikimedia.org, which until a few years ago freely provided Wikipedia database dumps to the research community. Since 2010, data releases have become much more difficult and of inconsistent quality, so we thus decided to use the last reliable corpus available.

With these restrictions in mind, we identified approximately 235 million editorial changes made by 22 million unique editors. Of these, three million were unique registered users, while the rest were anonymous editors represented only by their respective IP addresses.

The raw editorial data was processed with the WikiTrust parser (Adler et al. 2008), which analyzes each editorial intervention and calculates its "contribution delta" (d)—the amount of effective change and, by proxy, of effort invested in the change. As noted in Chap. 4, the d formula takes into account all three kinds of changes that could be recorded for a given intervention: additions, deletions, and changes to existing text. It aims to provide a moderated score that accounts for but also limits the impact of deletions and mere copy-paste changes when compared with additions of new content. In this respect, this formula represents a superior measurement strategy over previous attempts made in similar studies, which typically used mere character counts to estimate the significance of different contributions.

For reference, the contribution delta formula, which was also provided in Chap. 4, is as follows:

$$d(u,v) = \max(I, D) - 0.5 \min(I, D) + M, \qquad (5.1)$$

where I = total inserted text, D = total deleted text, and M = total relative change of position (a measure of text reordering).

The d score is used throughout this study as the main measure of user contributions. Ascertaining the exact quantity of material contributed by each individual user further allowed us to calculate three key parameters: the degree of system inequality, the presence or absence of a given individual in the contributing elite, and the strength of collaborative connections among fellow contributors.

First, system inequality was measured using an application of the entropy formula (Shannon and Weaver 1948):

$$H(X) = -K \sum_{m}^{i=1} p_i \log_2 p_i \qquad (5.2)$$

where K is a constant.

For each week, a normalized entropy value between 0 and 1 was calculated to ensure comparability across weeks. Normalized entropy was 1 when observed entropy equaled the maximum possible entropy level, with lower levels of entropy signifying a higher level of group structuration (see Appendix A for details). With this in mind, we measured group entropy at a weekly level, considering all contributions dating back to Wikipedia's launch, for a total of 495 weeks. (For instance, the first measurement used all contributions made in Wikipedia's first week, the second accounted for all revisions made in the first 2 weeks, and so forth.) These data are presented in Appendix A, Table A.2.

Second, to evaluate elite group stickiness, we measured the percentage of individuals present among the elites from one 5-week period to the next. The intervals for elite stickiness are longer than the ones for entropy because 1 week is not a sufficiently long period to distinguish elite editors from more casual contributors. The 5-week interval was determined empirically by observing the intervals that offered the best smoothing function. Additionally, for the sake of ensuring sufficient data to define an elite group, the first 3 weeks were dropped from this assessment, which left 98 measurement points. This data set is presented in Appendix A, Table A.3.

The last key parameter was the strength of coeditorial connections between editors. This measure was necessary in order to see if nonelite members' connections to members of the elite would impact their probability of being in the elite at a future time. To assess this final parameter, we constructed a graph of coeditorial connections for each week to summarize the intensity of implicit collaboration between all editors up to that point in time. Our measure of intensity used a gravitational model, where two editors who edited the same article in close succession and whose contributions were both substantial in nature held a stronger connection. This method is similar to the gravitational model for constructing economic, social, and organizational networks, pioneered by Isard (1954, 1975). Specifically, the formula is

$$F_{ij} = \frac{M_i M_j}{d_{ij}^2} \tag{5.3}$$

where M_i is the contribution delta of the revision made by the first editor, M_j is the contribution delta of the revision made by the first editor, and d_{ij} is the distance between the revisions, measured by consecutive edits.

If the same two editors coedited more than one article, or if they coedited the same article through multiple iterations, their total connection was deemed equal to the sum of all their connections across all articles (Britt 2011, 2013; Matei et al. 2015). Again, for details, see Appendix A.

5.2.2 Overall Analysis Strategy

In Chap. 3, we proposed five research questions to direct our investigation:

RQ1. Is there a distinct group of highly productive (elite) users on Wikipedia, and if so, to what extent do they dominate contributions to the project over time?
RQ2. What is the social mobility (or its inverse, elite "stickiness") of functional leaders on Wikipedia over time?
RQ3. Are there distinct growth phases determined by the emergence of functional leaders in Wikipedia's first 9 years of existence (2001–2010)?
RQ4. What is the relationship between global inequality, or social structuration, and elite stickiness?
RQ5. Is promotion to the elite group a function of prior interactions with elites?

We start by determining the contours of the contributing elite. If members of the elite are, as would be expected, one or several orders of magnitude more productive than the rest of the editors, then the distribution curve would follow a power law, with a definite inflection point. After ascertaining the presence of a power-law distribution at the global level with an inflection point, data for each week was similarly plotted, and the respective inflection points that differentiated the elite contributors from the rest during each individual week were determined as well. Finally, the weekly inequality level was calculated using entropy.

The percentage of users in the elite varied, but on average, this was essentially the same as that of the global level, at approximately 1%. For each 5-week period, we calculated the level of elite stickiness, and average levels of entropy for 5-week intervals were also calculated so that these measures could be aligned. Next, we used a stepwise segmented regression analysis to determine the discrete phases of inequality (structuration) for the entire period (2001–2010). For further details about this procedure, see Chap. 9 and Appendix C.

Given the longitudinal nature of the data, we also evaluated the presence and intensity of autoregressive processes in the two core variables, entropy and stickiness. An ACF (autocorrelation function) suggested a baseline predictive model and its directionality, which was tested using lagged regression (Shumway and Stoffer 2010).

Finally, we tested the probability that a current nonelite member would be present in the elite at a future point in time on the basis of his or her connections to members of the elite as well as the amount of past and present contributions made. The different factors predicting elite ascension were plotted onto the evolutionary phases predicted by Wenger (1998) to determine if recruitment processes changed along with those phases or were independent of them.

5.2.3 Descriptive Exploration

Over the study period (2001–2010), we observed 235,701,162 edits completed by 22,792,847 unique contributors. Of these, 19,680,637 users were anonymous, identified only by their unique IP addresses. The rest (3,112,210) were registered users who were logged into their respective accounts. As previously mentioned, logged-in users were the clear minority group, yet they contributed far more edits than the anonymous users—all told, those logged-in individuals were responsible for almost two-thirds (68%) of the observed revisions. Even more importantly, the top 1% of all contributors were responsible for 77% of the collaborative effort based upon the extent to which the text of articles was actually changed (i.e., the contribution delta). Logged-in users were also more important among elite users—within the top 1% of contributors, those who were logged in accounted for 82% of the elite group's collaborative effort.

The contribution quantity follows a power-law distribution (long tail, sharp head slope) with a definite inflection point defining the top 2–3% of contributors (Fig. 5.1). Thus, the "head" of the curve includes approximately the top 1% of all contributors, the "neck" the top 2–3%, and the tail the remaining 97%. We used a maximum likelihood estimation technique combined with bootstrapping using the poweRlaw library in R to determine that the observed distribution indeed belongs to the power law family ($p > 0.1$; see Clauset et al. 2009).

These findings indicate that the first part of research question 1 (RQ1), "Is there a distinct group of highly productive (elite) users on Wikipedia?" should be answered in the affirmative. Even more significantly, the majority of the contributions made

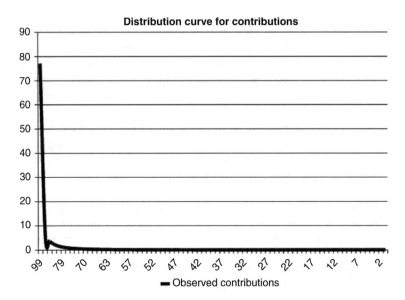

Fig. 5.1 Power-law distribution of contributions to Wikipedia, 2001–2010. X-axis: contributor percentile. Y-axis: percentage of contributions

by elite members belong to the logged-in users. The logged-in members of the top 1% contributor elite singlehandedly comprise a very significant share of contributions made to Wikipedia, measured by amount of effective work. This makes sense, as one can easily guess that logged-in users, who bother to create online names and identities to which their efforts are attributed, would be the most invested and motivated to enhance the long-term structuration of the project. As we can see, those logged-in elites thoroughly dominate the project, thereby answering the second part of RQ1 as well.

Furthermore, it is worth recognizing that IP addresses would be unreliable indicators for determining the unique identities of contributors over long periods of time, so any structuration analysis using that data would be unreliable as well. On the basis of this practical reality as well as our finding about the relative importance of registered users—even among the elites—we chose to focus the rest of our research on logged-in users, since they carry the weight of the project.

Therefore, the rest of the analysis focuses on the social structuration and elite temporal behavior among a subset of Wikipedia contributors, those users logged into registered accounts, as they are the most heavily invested in the project and contribute the most to it.

With that in mind, mapping the inequalities among editors in terms of both raw effort (amount of d contributed by the top logged-in users) and the entropy level of all contributions allows us to observe the trajectory of the structuration process as it developed on a week-by-week basis throughout the 9.5-year period (Fig. 5.2). For the sake of visual comparisons between variables, all entropy values shown in Fig. 5.2 have been multiplied by 100.

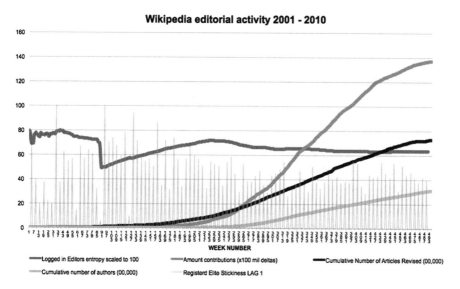

Fig. 5.2 Editorial dynamics for logged-in users from 2001 to 2010

On this chart, entropy has an interesting trajectory, although when we consider Wenger's (1998) perspective on organizational evolutionary phases it becomes quite understandable. After a period of turbulence spanning its first 92 weeks, the Wikipedia project quickly gained new editors and progressed through a period of increasing egalitarianism (at least in relative terms). However, in week 226, the project reached a minimum of relative inequality (and maximum entropy), and after that week, inequality steadily increased over time. During this period, entropy declined (and inequality correspondingly increased) by 12%, from 0.719 to 0.635. Importantly, this inequality continued to grow even as the numbers of articles, authors, and editorial interventions rapidly swelled.

Elite resilience (stickiness), represented as gray spikes marking the end points of our 5-week intervals, varied widely in the first half of the project, when participation was more chaotic and less structured. After week 290, following the decline in entropy and the corresponding increase in structural differentiation, it started increasing at a steady pace. Setting aside these variations, the simple answer to research question 2 (RQ2), "What is the social mobility (or its inverse, elite "stickiness") of functional leaders on Wikipedia over time?" is that on average, across the entire 9.5-year period, an individual who was a top contributor at a given point in time had a 40% probability of remaining in the top contributor group 5 weeks later. Twenty weeks later, that individual would have a 32% chance of still being a top contributor, and after 30 weeks, this figure would be at 28%.

5.2.4   Inferential Analysis

In addition to RQ1 and RQ2, both of which were broad, exploratory questions, we proposed three research questions that explore specific details pertaining to the relationships between elite stickiness, social structure, and organizational evolution.

RQ3. Are there distinct growth phases determined by the emergence of functional leaders in Wikipedia's first 9 years of existence (2001–2010)?

RQ4. What is the relationship between global inequality, or social structuration, and elite stickiness?

RQ5. Is promotion to the elite group a function of prior interactions with elites?

Research question 3 (RQ3) deals with a crucial issue regarding social media spaces. Are they evolutionary organisms? Can we detect discrete phases in their evolution? To address this matter, we adapted and extended segmented (piecewise) regression analysis. Typically, this type of analysis is used when, for different ranges of X values explaining a response variable Y, there are different intercepts, slopes, or higher-order terms separated by clear breakpoints. Segmented regression allows different models to be fit to different segments of the data. In the process, model boundaries will indicate the breakpoints in the explanatory data.

In a segmented regression model, one of the main explanatory variables is an indicator function, which acts as a dummy variable. According to Britt (2013), each indicator function corresponds to all data points following a given point in the data set. Thus, by employing a stepwise selection procedure, indicator functions may be added and removed from the model using a predetermined selection threshold, such as the common $\alpha = 0.15$ standard (Crawley 2007). A given data point is a statistically significant breakpoint whenever its indicator function has a statistically significant effect the model, and that breakpoint signifies that the data segment following it has a significantly different intercept, slope, or higher-order term than the segment preceding the breakpoint. In other words, there is a clearly discernible change in the model.

Our stepwise segmented regression analysis used the week number as the predictor variable to predict entropy. In other words, we modeled changes in entropy over time, so any breakpoints represented fundamental changes in the growth or stability of entropy. We identified several breakpoints, which are listed in Table 5.1[2]:

By comparing the intercepts occurring before and after each breakpoint (see the parameter estimates in Table 5.1), we notice that some shifts are very significant in magnitude (e.g., weeks 92, 204, and 250), while other breakpoints denote much smaller changes (such as weeks 42 and 335). More importantly, visually inspecting

---

[2] For the sake of simplicity, only the intercept terms are provided in Table 5.1. The complete regression model including all intercept, slope, and quadratic terms is provided in Appendix A. Likewise, in Chap. 9, we identify several other breakpoints denoting significant transition points in the evolution of the project. These emerged from analyzing four other dimensions of collaboration that are not discussed in this chapter. Again, a full description of the significance of all breakpoints is provided in Appendix A.

Table 5.1 Breakpoints in stepwise segmented regression model for entropy

	Estimate	Std. error	t Value	p-value	
Intercept	$3.408E^{-1}$	$1.034E^{-1}$	3.297	0.00105	
Intercept*I (week $\geq$7)	$-2.377E^{0}$	$2.872E^{-1}$	-8.277	$1.34E^{-15}$	*
Intercept*I (week $\geq$10)	$2.825E^{0}$	$2.679E^{-1}$	10.542	$<2E^{-16}$	*
Intercept*I (week $\geq$42)	$1.559E^{-1}$	$1.227E^{-2}$	12.711	$<2E^{-16}$	*
Intercept*I (week $\geq$92)	$-6.323E^{-1}$	$1.358E^{-2}$	-46.574	$<2E^{-16}$	*
Intercept*I (week $\geq$93)	$-8.551E^{-2}$	$3.586E^{-3}$	-23.843	$<2E^{-16}$	*
Intercept*I (week $\geq$204)	$-1.623E^{0}$	$1.652E^{-1}$	-9.822	$<2E^{-16}$	*
Intercept*I (week $\geq$250)	$3.325E^{0}$	$1.753E^{-1}$	18.974	$<2E^{-16}$	*
Intercept*I (week $\geq$335)	$-8.865E^{-1}$	$6.376E^{-2}$	-13.905	$< 2E^{-16}$	*

*Denotes statistically significant variables after a Holm-Bonferroni correction ($\alpha = 0.05$)

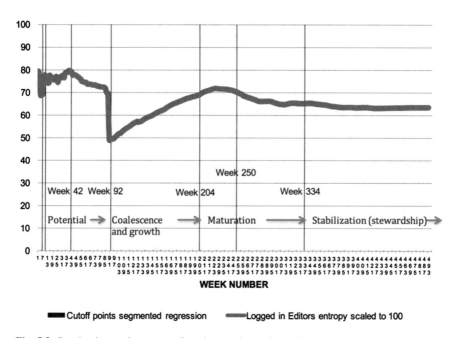

Fig. 5.3 Breakpoints and corresponding change phases detected via stepwise segmented regression analysis

the curve (Fig. 5.3) and comparing it with the parameter estimates from Table 5.1 allow us to identify four main phases spanning weeks 0–91, weeks 92–203, weeks 204–334, and week 335 to the end of the data set.

These periods map well onto Wenger's (1998) phases of organizational evolution. The first phase, potential, is characterized by wide swings in behavioral norms and expectations—including leadership dynamics—which fits the erratic vacillations observed during the first 91 weeks. An approximate midpoint at week 42 marks a maximum of entropy for the entire period. The second period, coalescence, is theorized to be one of rapid growth, characterized by a sharper evolutionary slope, as it is in our case from weeks 92 to 203. Notably, the abrupt drop in entropy at week 92 is the product of the introduction of the first bot (rambot), which added a large number of articles in a very short period of time. This occurred early in the life of the project, and its contributions were, in time, overshadowed by the larger-scale organic growth of the project. Nonetheless, this intervention is significant considering its impact on the collaborative effort at the time, so it will be discussed at length in Chap. 9.

The last two periods that Wenger (1998) described are those of maturation and stabilization or stewardship. In our context, we may observe that the entropy curve shown in Fig. 5.3 exhibits a distinct "cooling off" (maturation) phase between weeks 204 and 334 and a "leveling off" (stabilization) phase from week 335 to the end of the data set.

This alignment of our breakpoints onto an existing evolutionary model provides a clear answer to RQ3, as it demonstrates that the social structuration of Wikipedia included discrete and well-defined evolutionary phases that followed predictable steps of growth and stabilization. This strengthens the claim that Wikipedia, as a social space of collaboration, followed an evolutionary path, which by definition involved structural differentiation. With that said, we still require a more complete demonstration of the phase approach for revealing collaborative structures and, more importantly, a concrete articulation of the identity of those structures. This will be provided in Part II of this volume, especially in Chaps. 8 and 9.

Moving on, research question 4 (RQ4) is probably the most important of our study, as it addresses the crucial relationship between elite resilience (stickiness) and organizational structuration (entropy). In other words, we ask whether processes that occur at local (intra-elite) levels interact with global structuration processes.

To answer this question, we analyzed the relationship between two variables, elite resilience (stickiness) and entropy, using a lagged regression approach. This procedure is used to discern the effect of past values of the predictor variable (entropy) on future changes in the variable of interest (stickiness). Importantly, we focused on the periods of growth and maturity occurring from week 92 onward. The reasons for this are twofold. First, as we will explain in Chaps. 9 and 10, up to week 92, Wikipedia went through two major socio-technological changes, including a new database structure and a massive injection of content that was automatically generated. Given the small size of the project (hundreds and thousands of articles and authors compared to the millions by the stabilization phase), these early changes made the data very noisy. Second, our focus is on detecting the relationship between stickiness and entropy in the context of adhocracy, which starts, as we will show in Chaps. 9 and 10, after week 92.

Given the time series nature of the data, we analyzed the cross-correlation function (CCF) using the *ts* library in R to determine what leads the process: structuration or elite stickiness. We regressed the stickiness of the logged-in editors on lagged values of entropy and found that changes in entropy lead to changes in stickiness, not vice versa. Figure 5.4 shows that, with entropy as the independent variable to

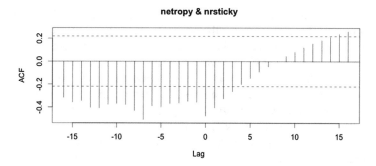

Fig. 5.4 ACF analysis of the cross-correlation between entropy and elite stickiness (resilience) on Wikipedia

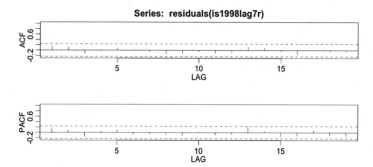

Fig. 5.5 ACF and PACF of residuals for the cross-correlation between entropy and elite stickiness (resilience) on Wikipedia

predict stickiness, the highest lag is the seventh negative one. Negative lags indicate that changes in the X variable, entropy, lead and potentially cause those for the Y variable, stickiness. Since elite stickiness was measured every 5 weeks, this means that changes in entropy have a maximum and statistically significant effect 30–35 weeks later. Furthermore, the correlation value for lag 7 is negative, at −0.5. In conceptual terms, as the level of structuration increases—which we observe as a decrease in entropy—the likelihood that current members of the elite will still be in the elite at a future point in time also increases. Thus, increasing structuration results in increasing elite stickiness.

Lagged regression analysis of the effect of lag 7 of entropy on stickiness, controlling for unlagged values of entropy, yielded a negative (unstandardized $b = -2.21$, standardized $\beta = -0.94$) and highly significant ($p < 0.001$) effect. Furthermore, almost half of the variance in stickiness is explained by present and past values of entropy ($R^2 = 0.47$). Our ACF analysis of the residuals indicates that the autocorrelation function is not systemic, suggesting a good model fit (see Fig. 5.5).

All told, then, the answer to RQ4 is that there is a significant relationship between social structuration and elite resilience, with the former preceding the latter. Structuration precedes elite consolidation, not the other way around.

Research question 5 (RQ5), the last at this stage of our discussion, addresses the mechanism by which members of the elite are promoted in the period of maturity. Two models of ascension may be proposed. The first model presumes that advancement is osmotic, as new members rise among the most productive contributors solely through their own efforts, with no support from those already in the elite. The second instead assumes that interacting with the elite members by collaborating with them on specific articles in close succession constitutes a process of implicit acculturation and social training. While the latter model does not assume collusion, it does assume that collaborative work increases the chances of interaction and the acceptance of one's own contributions. This may motivate nonelite users, whose increased productivity would then lead them to an elite status.

We tested both models using logistic regression. The dependent variable was whether or not current elite and nonelite members were present in the elite group in the future (specifically, during the following 5-week interval), while the key independent variable was the strength of one's existing connection to elite members in the current 5-week interval. Elite membership was defined in terms of the location of one's own contribution quantity, whether to the right (tail, low contribution) or left (head, high contribution) of the inflection point in the power-law distribution of contributions in each 5-week period (see Fig. 5.1).

The following predictor variables were also introduced into the model:

- Contribution quantity in the current 5-week period (lag 1) and the immediately previous period (lag 2)
- Number of collaborative connections to elite and nonelite members in the current period (lag 1) and the previous one (lag 2)
- Weight of connections to elite and nonelite members in the current period (lag 1) and the previous one (lag 2)

The number of connections was operationalized as the number of times a given pair of individuals coedited any article, while "weight" was the sum of the value of the connections formed between the pair as specified by the gravitational model (Eq. 5.3).

In order to handle the massive amount of available data, we used a random sample of up to 1,000 individuals from the elite and 1,000 from the nonelite for each 5-week interval. With an effect size of 1,000 contribution delta units and $\alpha = 0.05$, the power of the random sample was 0.994, a high value that justifies a sample-based analysis. However, this approach compelled us to skip the first 100 weeks within the data set, during which the total number of contributors for many 5-week periods was lower than 2,000.

A logistic regression approach demands that the dependent variable be transformed into a logit odds ratio, $\log \frac{p}{1-p}$, such that the regression model then

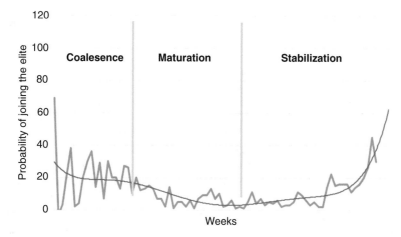

Fig. 5.6 Odds ratios and polynomial fitted trend line for logistic regressions predicting future presence in the elite as a function of a user's own contribution quantity in the present period, adjusted by a factor of 100,000

becomes $\log \dfrac{p}{1-p} = \beta_0 + \beta_1 X_1 + \beta_2 X_2 + \cdots + \beta_8 X_8$. Additionally, an iteratively reweighted least squares method was applied to maximize the likelihood function, so in our case, the higher the odds ratio for the parameter estimate, the more likely that a user belonged to the elite group.

Upon fitting this model, we concluded that one's connection to elite members, whether based upon number of coedits or the sum of connection weights, did not significantly predict future presence among the elite. The only significant and consistent predictor was the amount of contribution, or contribution delta, for all revisions made by the user in the preceding period. In other words, the only statistically significant predictor of a given user's presence in the elite group was the amount of content that he or she contributed in the immediately preceding 5-week period.

Figure 5.6 shows the odds ratios for presence in the elite group based upon one's past contributions. Because the contribution metric is represented in small increments, representing minute changes for each editorial intervention, we multiplied each odds ratio by 100,000. This allows us to see the effect of an individual contributing an additional 100,000 units of effort, or contribution delta, on his or her chances of subsequently being in the elite group. The effect is estimated in percentages. That is, we estimate the percent increase in the probability of being in the elite in the next 5-week interval, if a member of the nonelite contributes an extra 100,000 contribution units in the present 5-week interval. Figure 5.6 indicates that after week 101, the effect of one's past contributions on the probability of joining the elite is almost always positive—in other words, a greater amount of effort always increases the likelihood that a given user will subsequently ascend to the elite group. As an example, an increase of 100,000 in a given user's cumulative contribution delta value between weeks 195 and 199 would result in a 10% increase in his or her probability of entering (or remaining in) the elite in the week 200–204 interval.

As previously noted, this analysis specifically targets collaborative dynamics as they developed after the potential phase (Wenger 1998). To that end, week 101, the starting point for this analysis, is also very close to week 92, the boundary between the coalescence and stabilization phases identified in the answer to RQ3.

Overall, we observe that the impact of hard work on one's promotion to elite status was greater at the beginning and at the end of the study period, corresponding to the coalescence (weeks 101–204) and stabilization (weeks 335–495) intervals. Between weeks 101 and 334, corresponding with the rapid growth and maturation period previously identified for RQ3, the log odds declined, indicating a weakening effect of one's past contributions on a future rise to the elite group.

To put it in more tangible terms, during the period spanning weeks 101–204, an additional 100,000 contribution units generated, on average, an increase of 20% in the probability of joining the elite. In the 205–334 interval, the increase in the probability of joining the elite from each additional 100,000 units of contribution declined to an average of 6%. In the last period, however, after week 335, the impact of a user's prior contributions increased once more. Overall, for the entire period, the average increase in the probability of joining the elite due to each 100,000 units was 11%. And, during the period from weeks 435 through 495, the average probability was even higher, returning to the 20% value recorded at the beginning of the study period. Further, the last statistically significant log odds ratio, obtained for weeks 490–494, showed a jump of its own, indicating that 100,000 in contribution delta would result in a 45% increase in the likelihood of reaching the elite ranks.

5.3 Looking Back, Looking Forward

The findings presented so far suggest that deep contribution imbalances in social media projects with deep contribution imbalances are, in fact, evidence of an evolutionary process. Some members will naturally grow to play high contributor roles, and the system will evolve with them in distinct phases of social structuration: potential/incubation, coalescence, maturation, and stewardship/stability. Thus, social media projects like Wikipedia are not a mere "wise crowd" in which individuals perform similar functions with minimal social coordination in approximately the same proportion and using roughly the same approach. Social structure and a stable elite are also essential to the process. In fact, social structuration leads to "stickier" elites, with structuration explaining almost half of the variance in elite stickiness.

Most importantly, we uncovered an apparent causal relationship, as elite stickiness increases only *after* entropy declines. This means that elites "gel" in their positions after the project as a whole becomes top-heavy. The top functional elite leaders are organically encouraged by the success of their content to stay with the project for longer periods of time. Their presence on the project becomes very important, as their ceaseless activity becomes increasingly visible and harder to replace due to its sheer mass. In addition, functional leaders who have a greater vested interest in the project are more likely to return to the site in order to ensure the continuity of their efforts. Thus, the likelihood that their content will be preserved and even increase in mass will be enhanced accordingly.

Our findings contradict alternative explanations that claim that contributors freely rotate into and out of elite roles as in a game of musical chairs—thus making any differentiation between contributors even out over time—and, therefore, that any observed inequality would essentially disappear if we took elite turnover into account. For this to be true, elite stickiness would have to be almost zero. What we found, instead, is that stickiness is relatively high, at 40% across the duration of the study.

Finally, we determined that an osmotic model of social promotion based on individual effort was more useful for explaining one's future presence in the elite than coeditorial connections to existing elite members. However, this relationship was not linear. In particular, during the period of coalescence, hard work did not always pay off, or at least not to a substantial degree. This finding might be the result of a rapid increase in the number of contributors during that phase. After all, when many new contributors join a project at once, the "social dividend," or probability of rising to the elite, declines as competition rapidly intensifies. Only when the project stabilizes does the "social dividend" of hard work begin to pay off for those contributors who persisted in their efforts. Toward the end of the study period, for each 100,000 units contributed during a 5-week period in terms of delta scores, the likelihood of joining the elite (or staying in it) increased by 45%. Taken together, these findings exemplify the stabilization and "stewardship" phase that we would expect from a fully mature collaborative project.

These findings are essential for understanding the formation of social groups online, especially on social media, since they rely on an intact data set of an entire social medium over almost a decade of activity. Still, in looking forward to Part II of this volume, it is important to recognize that the mechanisms described thus far are rather basic. They only explain the incubation, coalescence, growth, and stabilization of social media collaborative groups in terms of general trends, referring loosely to concepts like adhocracy as omnibus terms to explain the phenomena. However, complex projects such as Wikipedia span a variety of work and interactional contexts, some of which may be more traditional than newer ideas like adhocracy would suggest, and many of which demand more nuanced explanations for the phenomena at hand.

As we draw Part I of this volume to a close, we still need a clearer understanding of the collaborative dynamics behind the contributions that we observed and the organizational phenomena that drive them. This includes, crucially, the ways in which specific organizational configurations develop as well as how they are entangled with the larger evolutionary process that we have observed thus far.

In the next two chapters, we will propose a more sophisticated interpretation of the evolutionary process and a more detailed description and interpretation of the adhocracy concept to which we have referred thus far. We have acknowledged the conceptual importance of adhocracy as a broad process, but at the same time, when it is rigorously applied and defined, it also represents a very specific type of organizational arrangement. For the sake of this second perspective on adhocracy, we need to consider the multiple facets of the collaborative process and outline the main motors and configurations of organizational change and transformation. This approach requires the addition of new dimensions to our research, including a more complex organizational change model, which will facilitate a more nuanced exploration of online collaborative evolution. These objectives will be addressed at length in Part II.

References

Adler BT, de Alfaro L, Pye I, Raman V (2008) Measuring author contributions to the Wikipedia. http://citeseerx.ist.psu.edu/viewdoc/download?doi=10.1.1.141.1107&rep=rep1&type=pdf. Accessed 8 Feb 2017

Britt BC (2011) System-level motivating factors for collaboration on Wikipedia: a longitudinal network analysis. Thesis, Purdue University

Britt BC (2013) Evolution and revolution of organizational configurations on Wikipedia: a longitudinal network analysis. Dissertation, Purdue University

Clauset A, Shalizi CR, Newman MEJ (2009) Power-law distributions in empirical data. SIAM Rev 51:661–703

Crawley MJ (2007) The R book. Wiley, West Sussex

Isard W (1954) Location theory and trade theory: short-run analysis. Q J Econ 68:305–322

Isard W (1975) A simple rationale for gravity model type behavior. Pap Reg Sci 35(1):25–30

Kriplean T, Beschastnikh I, McDonald DW, Golder SA (2007) Community, consensus, coercion, control: cs* w or how policy mediates mass participation. In: Proceedings of the 2007 international ACM conference on Supporting group work, p 167–176

Kriplean T, Beschastnikh I, McDonald DW (2008) Articulations of Wikiwork: uncovering valued work in Wikipedia through Barnstars. In: Proceedings of the 2008 ACM conference on computer supported cooperative work. ACM, New York, pp 47–56

Matei SA, Bertino E, Zhu M, Liu C, Si L, Britt BC (2015) A research agenda for the study of entropic social structural evolution, functional roles, adhocratic leadership styles, and credibility in online organizations and knowledge markets. In: Bertino E, Matei SA (eds) Roles, trust, and reputation in social media knowledge markets: theory and methods. Springer, New York, pp 3–33

Shannon CE, Weaver W (1948) The mathematical theory of communication. University of Illinois Press, Urbana

Shumway RH, Stoffer DS (2010) Time series analysis and its applications: with R examples, 3rd edn. Springer, New York

Wenger E (1998) Communities of practice: learning, meaning, and identity. Cambridge University Press, Cambridge

Part II
Configurational Change Phases and Motors in Online Collaboration

Chapter 6
The Foundations of a Theoretical Model for Organizational Configurations and Change in Online Collaborative Processes

6.1 Introduction

So far, we have presented an evolutionary model for the development and structuration of collaborative groups. We have shown that as collaborative entropy on Wikipedia decreased, a stable elite took shape, with the former process driving the latter. We have also presented considerable evidence for a process of "adhocratic" change in elite behavior and for longitudinal development with distinct phases of growth and maturation. However, this analysis captures a global process at a high level of abstraction. As such, it does not specify the possible mechanisms by which social structuration, which we measured through entropy, actually takes place.

In the text that follows, we will develop a more complex theoretical model, which includes a set of discrete organizational configurations that may be found in the life cycle of a collaborative organization. We will also propose some "motors," or transformative factors, that drive the process of evolution and structuration. This model is grounded in organizational communication and organizational behavior research, while the empirical testing of these factors is rooted in the more specialized field of computational social science.

The remainder of this chapter introduces the long-standing gap in the literature at the intersection of organizational configurations and organizational change. Chapter 7 delves further into the concept of organizational configurations, which has an extensive history in the management literature. Then, in Chap. 8, we unite the domain of organizational configurations with that of organizational change motors, thereby yielding a synthesized theoretical framework that may be used to explain the entire process by which an organization's form changes over time. We then employ this framework in Chap. 9 to determine the mechanisms of online social structuration, with the broader ramifications of our findings detailed in Chap. 10.

© Springer International Publishing AG 2017
S.A. Matei, B.C. Britt, *Structural Differentiation in Social Media*, Lecture Notes in Social Networks, DOI 10.1007/978-3-319-64425-7_6

6.2   Toward a Theoretical Explanatory Model of Organizational Change in Online Collaborative Structures

Our model is rooted in the theoretical perspectives offered by research on organizational configurations and organizational change. In this context, "configurations" refer to patterned and recurring interactions and role allocations, with the concept speaking in a more direct way about the processes of social organization and differentiation that were introduced in the previous chapter. Organizational change research, in turn, deals with why and how large groups of individual communicators develop in particular ways or along specific trajectories over time.

Configuration research has proven especially useful for classifying organizations into meaningful typologies, while change scholars have emphasized the long-term development of organizations as opposed to the examination of mere snapshots in time. Yet there has been minimal crossover between these two areas, with a particular lack of research devoted to organic changes to organizational configurations emerging from within the organization itself. More importantly, the literature does not offer a clear connection between these areas, nor does it address the issue of social structuration in voluntary online organizations in any fashion.

With that in mind, this chapter begins the process of intertwining the domains of organizational configurations and organizational change into a synthesized framework of configurational change motors. The goal is to articulate a comprehensive theoretical model that can bring the conversation about social structuration, covered in the previous chapters within a very broad and abstract terrain, to a more tangible and precisely specified reality. The first step is to identify specific types of system-level structuration, which in keeping with existing literature (e.g., Mintzberg 1979, 1989) we will call "organizational configurations." Second, we will discuss the necessity and sufficiency of the configurations identified to date. Third, we will derive methods of translating the conceptual definitions of the configurations into specific operationalizable concepts. Fourth, we will lay the groundwork for the specific mechanisms by which configurations change and succeed each other over time (Van de Ven and Poole 1995), which will be further developed in Chap. 7.

Our theoretical model extends, on the one hand, the framework of Mintzberg's (1979, 1989) Configurational Array, which integrates many otherwise fragmented theories into one simple explanatory mechanism grounded in organizational processes and structures that are relatively easy to measure. And, on the other hand, we also build upon Van de Ven and Poole's (1995) system of organizational "motors," which is arguably the single most comprehensive synthesis of existing theories on the internal dynamics and environmental forces that propel organizational change.

In brief, this chapter aims to connect our very macro-theoretical proposition about structural differentiation, which was delineated in very broad terms through its definition and our measurement of it through entropy, with meso- and micro-level theories that detail and explain specific organizational configurations as well as the transition processes between them. All told, we explain, in especially tangible terms, how social structuration emerges and develops over time.

It bears noting that our discussion about organizational configurations distinguishes between configurational archetypes and an organization's present configuration. A configurational archetype is an ideal type, to use Weber's (1946) theoretical concept: a possible and greatly simplified organizational pattern that may be found in an organization at a certain point. These archetypes stand as poles on a spectrum, representing extreme cases in order to provide a frame of reference for understanding. Present configurations, on the other hand, are observable and measurable patterns that may closely resemble one configurational archetype or fall between two or more archetypal patterns. The distinction is that between the concept itself and the observable instantiation of the concept. It is analogous to the distinction between the concept of "organization"—which denotes the idea of a group of people who interact with each other according to certain rules, roles, values, and intended goals—versus the various groups in the world that we identify as "organizations."

Our model utilizes a few common organizational constructs to assess the configurational archetype toward which the present form of an organization is moving at any given moment, which thus facilitates the identification of key evolutionary trends and revolutionary moments over time. Once these critical change periods and instances have been identified, we may then explore a variety of internal and external factors in order to assess how they push the organization from one configurational archetype toward another. Thus, the configurations are linked with the change motors through a set of conventional organizational measures as well as an interpretive assessment of the motors that contributed to the observed changes, using the dichotomy of evolutionary and revolutionary change as a conduit for the theoretical integration. This ultimately serves to unite these traditionally disparate research areas and permit the assessment of configurational change motors, including those that we may readily observe within the increasingly common member-defined online organizational configurations throughout the modern world.

6.3 Configurational Change: Critical Theoretical Gap and Bridges

Over the past few decades, a great deal of research has been conducted on classifying organizations into meaningful configurations based on the ways in which they operate, most commonly focusing on their respective structures, the managerial strategies that undergird them, and the core missions that they serve. Numerous researchers have devoted substantial attention to the kinds of configurations that are commonly implemented within organizations and the consequences of those managerial implementations (e.g., Birkinshaw and Lingblad 2005; Davidow and Malone 1992; Dess et al. 1997; Handy 1989; Heckscher 1994; Kang et al. 2007; Kaufman et al. 2000; Korunka et al. 2003; Kotha and Vadlamani 1995; Lucas 1996; Menguc and Auh 2010; Miles and Snow 1986; Mowshowitz 1994; Mueller et al. 2003; Nair and Filer 2003; Payne 2006; Pegels et al. 2000; Rauch et al. 2004;

Spanos et al. 2004; Stabell and Fjeldstad 1998; Stordeur and D'Hoore 2007; Veliyath and Shortell 1993; Wiklund and Shepherd 2005; Youndt and Snell 2004). Together, these scholars have generated, for better or worse, an ever-increasing array of organizational characteristics and configurations that describe the particular systems they have observed across a range of contexts, along with just as many explanatory mechanisms for how and why those particular configurations developed (see Lammers 1988; McPhee and Poole 2001; Winter and Taylor 1996).

An equally rich research tradition can be found in the area of organizational change. Scholars vary in their area of interest within the organizational change literature, with foci running the gamut from stage theories featuring slower evolutionary changes punctuated by instantaneous revolutionary changes (e.g., Auzair and Langfield-Smith 2005; Flynn and Forman 2001; Gersick 1991; Greiner 1972; Hwang and Park 2006; Lester et al. 2003; Levy and Merry 1986; Meyer et al. 1993) to integrative concepts like organizational ecology (e.g., Becker 2007; Grote and Lang 2003; Monge et al. 2011) as well as the particular forces and motivations behind organizational transformations (e.g., Chakravarthy and Lorange 1991; Flavell 1982; Hegel 1812; Kimberly and Miles 1980; Matei et al. 2010).

Individually, both organizational configurations and organizational change have received substantial attention. But by and large, very little work has been devoted toward synthesizing these two areas in order to better study organizational development over time. Astley and Van de Ven (1983) pointed out the need for more research on configurational change 30 years ago, and in a more recent review of the configurations literature, Short et al. (2008) reiterated the problem that "...change and stability have not been major foci in recent years" (1067). Even less effort has been devoted to connecting organizational configurations and organizational change research to the much broader issue of structural differentiation at large, which further highlights the persistent gap in the literature.

With that said, it is clear that some scholars of organizational configurations are at least moving past the notion of the configuration as a static entity by studying the development of organizations over time. In particular, many researchers have devoted their attention to processes that extend or transform formally defined organizational structures (e.g., Carley and Svoboda 1996; Chan 2002; Cross et al. 2001; de Toni and Nonino 2010; Han 1996; Kratzer et al. 2008; Lee et al. 2002; Macintosh and Maclean 1999; Markus et al. 2002; Peng et al. 2004; Pullman and Dillard 2010; Schonpflug and Luer 2013; Shi 2010; Škerlavaj and Dimovski 2013; Uhl-Bien 2006), although few of these researchers have used their findings either to define new configurations or to classify organizations based on existing configurations. Furthermore, much of this work has only considered the initial emergence of such alternative structures or has treated them as static entities to merely be contrasted with formally defined organizational configurations—such as the informal communication that occurs outside the bounds of an official organizational chart—so the development of those emergent structures has remained largely unexplored. Nonetheless, these studies at least serve as a first step toward understanding the longitudinal development of emergent configurations and toward connecting specific configurations to stages, like those described earlier in this volume, in a more conceptual structural differentiation process.

As a whole, insufficient work has connected organizational configuration theory with existing theories on organizational change as it occurs over time. Siggelkow (2002) and Tyrrall and Parker (2005) serve as notable exceptions, but these are isolated examples that, importantly, remain confined to institutionally imposed configurations. By and large, few scholars have considered the types of change that organizational configurations undergo or the forces that drive those changes, particularly within the realm of emergent configurations.

This inattention may be due in part to overlapping and inconsistent terminology use. As Short et al. (2008) explained, configurational research has been conducted under many different names, and definitions of several key terms have varied significantly among scholars. Most of those differences arose between scholars from different disciplines whose respective literatures rarely intersect, highlighting the need for scholars to bridge disciplinary gaps and connect useful frameworks and theories with one another.

Further, some scholars have employed the term "configuration" itself toward other ends. For instance, "network configuration" refers to structural elements embedded within a larger network, from micro-level relations such as simple dyads and triads to higher-level tendencies like reciprocity and transitivity (Robins et al. 2007; Shumate and Palazzolo 2010; Snijders et al. 2006). Since organizational structures themselves are often a key criterion for classifying organizations into configurations and network configurations offer a natural (if underused) means of measuring those structures, this subtle terminology conflict is especially problematic for organizational configuration researchers. In fact, network configurations have themselves been connected with many different change theories, of which the most notable may be evolutionary and ecology theories (see Monge et al. 2011 for a review). Since the key term "configurations" has already been extensively used in the intertwining of network configurations with well-established change theories, that literature inadvertently masks the lack of a similar connection between organizational configurations and theories that would explain changes between them, as well as the related gap between the ideas of structural differentiation and evolution.

This literature gap is especially salient in the context of online voluntary organizations, such as sites like Wikipedia that were formed from user-generated content. After all, one of the very few areas in which organizational configuration research has focused on configurational changes has been the acknowledgement that emergent communication systems and organizational structures can arise from formally defined configurations, and even then, too little work has been done on the long-term change and growth of those emergent structures. Even less attention has been devoted to those organizations with no initial formal structure, whose configurations are defined from the very beginning solely by the actions of their members.

This particular gap is, in large part, a consequence of circumstance. Two decades ago, few could have imagined an organization that could (a) accumulate enough members to be defined in a more interesting and nuanced way than "team" or "small group," (b) survive long enough to grow and transform in any meaningful manner, and (c) do so with neither an external incentive system nor an organizational chart explicitly defined by a managerial system. Yet, with the advent of the Internet and of online collaborative platforms, it has become clear that organizations can emerge with little more than a name and an objective for members to fulfill and ultimately blossom into massive communities on par with the largest, most active and successful corporations in the world.

It is worth recognizing that developing an organizational configuration in this way, with minimal boundaries and resources for members, may be a viable approach for some offline organizations as well, even if online communities served as the first examples. With that said, the Internet offers an inexpensive space for communication and collaboration at any location and any time of day, reducing the risk for those who wish to undertake such an endeavor and increasing access to free (or at least cheap) labor. As such, it is obvious that the online domain does make Wikipedia-like efforts more feasible than they would be through face-to-face interactions alone.

6.4   Conclusion

As a whole, we have observed that it is possible for organizational configurations to be defined by their members from the very beginning, in contrast with other emergent configurations that instead merely adapt or escape an established formal structure. The process by which such a freely constructed configuration matures and grows, being actively defined and redefined by its members as the organization itself moves forward, is an entirely different dynamic that itself demands attention. As such, in addition to uniting the domains of organizational configurations and organizational change, this chapter also serves as an early endeavor into the study of configurations which do not emerge from institutionalized structures but which are initially formed and subsequently shaped only by the individuals embedded within them.

While a limited number of analyses have showcased how particular organizations move from one configuration to another, including those emergent structures that extend from formal hierarchies, not enough work has targeted changes that occur over long periods of time and the particular reasons why organizational configurations change in the ways that we have observed. Fewer studies still have connected case-specific causes for configurational changes with an established, generalizable theoretical framework that would explain long-term changes across organizations and contexts.

Our work, then, will intertwine these distinct frameworks. We delineate a set of key metrics that can be used to identify both periods of gradual evolution and moments of dramatic revolution in online and offline organizations alike, then extend this measurement approach to show how we can determine the ways in which an organizational configuration developed over time, and finally demonstrate a procedure to discern the motors that caused the configurational changes that have been identified. These constitute the critical objectives that we will address in the remainder of Part II.

References

Astley WG, Van de Ven AH (1983) Central perspectives and debates in organizational theory. Adm Sci Q 28:245–273

Auzair S, Langfield-Smith K (2005) The effect of service process type, business strategy and life cycle stage on bureaucratic MCS in service organizations. Manag Account Res 16:399–421

Becker F (2007) Organizational ecology and knowledge networks. Calif Manag Rev 49(2):1–20

Birkinshaw J, Lingblad M (2005) Intrafirm competition and charter evolution in the multibusiness firm. Organ Sci 16(6):674–686

Carley KM, Svoboda DM (1996) Modeling organizational adaptation as a simulated annealing process. Sociol Methods Res 25(1):138–168

Chakravarthy BS, Lorange P (1991) Managing the strategy process. Prentice Hall, Englewood Cliffs

Chan YE (2002) Why haven't we mastered alignment? The importance of the informal organization structure. MIS Q Exec 1(2):97–112

Cross R, Parker A, Prusak L, Borgatti SP (2001) Knowing what we know: supporting knowledge creation and sharing in social networks. Organ Dyn 30(2):100–120

Davidow W, Malone M (1992) Virtual corporation. Forbes 150(13):102–108

Dess GG, Lumpkin GT, Covin JG (1997) Entrepreneurial strategy making and firm performance: tests of contingency and configurational models. Strateg Manag J 18(9):677–695

Flavell JH (1982) Structures, stages, and sequences in cognitive development. In: Collins WA (ed) The concept of development: the Minnesota symposia on child psychology. Lawrence Erlbaum Associates, Hillsdale, pp 1–28

Flynn D, Forman A (2001) Life cycles of new venture organizations: different factors affecting performance. J Dev Entrep 6(1):41–58

Gersick CJ (1991) Revolutionary change theories: a multilevel exploration of the punctuated equilibrium paradigm. Acad Manag Rev 16:10–36

Greiner LE (1972) Evolution and revolution as organizations grow. Harv Bus Rev 50(4):37–46

Grote JR, Lang A (2003) Europeanization and organizational change in national trade associations: an organizational ecology perspective. In: Featherstone K, Radaelli CM (eds) The politics of Europeanization. Oxford University Press, Oxford, pp 225–254

Han SK (1996) Structuring relations in on-the-job networks. Soc Networks 18(1):47–67

Handy C (1989) The age of unreason. Harvard Business School Press, Boston

Heckscher C (1994) Defining the post-bureaucratic type. In: Heckscher C, Donnellon A (eds) The post-bureaucratic organization: new perspectives on organizational change. Sage, Thousand Oaks, pp 14–62

Hegel GWF (1812) Wissenschaft der logik. Johann Leonhard Schrag, Nürnberg. English edition: Hegel GWF (1998) Hegel's science of logic (trans: Miller AV). Prometheus Books: Amherst

Hwang YS, Park SH (2006) The evolution of alliance formation in biotech firms: an organizational life cycle framework. Manag Dyn 14(4):40–54

Kang SC, Morris SS, Snell SA (2007) Relational archetypes, organizational learning, and value creation: extending the human resource architecture. Acad Manag Rev 32:236–256

Kaufman A, Wood CH, Theyel G (2000) Collaboration and technology linkages: a strategic supplier typology. Strateg Manag J 21(6):649–663

Kimberly J, Miles R (1980) The organizational life cycle. Jossey-Bass, San Francisco

Korunka C, Frank H, Lueger M, Mugier J (2003) Context of resources, environment, and the startup process—a configurational approach. Entrep Theory Pract 28:23–42

Kotha S, Vadlamani BL (1995) Assessing generic strategies: an empirical investigation of two competing typologies in discrete manufacturing industries. Strateg Manag J 16(1):75–83

Kratzer J, Gemünden HG, Lettl C (2008) Balancing creativity and time efficiency in multi-team R&D projects: the alignment of formal and informal networks. R&D Manag 38:538–549

Lammers CJ (1988) Transience and persistence of ideal types in organization theory. Res Sociol Organ 6:205–224

Lee J, Lee K, Rho S (2002) An evolutionary perspective on strategic group emergence: a genetic algorithm-based model. Strateg Manag J 23(8):727–746

Lester DL, Parnell JA, Carraher S (2003) Organizational life cycle: a five-stage empirical scale. Int J Organ Anal 11:339–354

Levy A, Merry U (1986) Organizational transformation: approaches, strategies, theories. Praeger Publishers, New York

Lucas HC (1996) The T-form organization. Jossey-Bass, San Francisco

Macintosh R, Maclean D (1999) Conditioned emergence: a dissipative structures approach to transformation. Strateg Manag J 20:297–316

Markus ML, Majchrzak A, Gasser L (2002) A design theory for systems that support emergent knowledge processes. MIS Q 26(3):179–212

Matei SA, Bruno RJ, Morris P (2010) Visible effort: a social entropy methodology for managing computer-mediated collaborative learning. Paper presented at the Global Communication Forum, Shanghai, 29–30 Sept 2010

McPhee RD, Poole MS (2001) Organizational structures and configurations. In: Jablin FM, Putnam LL (eds) The new handbook of organizational communication: advances in theory, research, and methods. Sage, Thousand Oaks, pp 503–543

Menguc B, Auh S (2010) Development and return on execution of product innovation capabilities: the role of organizational structure. Ind Mark Manag 39:820–831

Meyer AD, Tsui AS, Hinings CR (1993) Configurational approaches to organizational analysis. Acad Manag J 36(6):1175–1195

Miles RE, Snow CC (1986) Organizations: new concepts for new forms. Calif Manag Rev 28:62–73

Mintzberg H (1979) The structuring of organizations: a synthesis of the research. Prentice Hall, Englewood Cliffs

Mintzberg H (1989) Mintzberg on management: inside our strategic world of organizations. Free Press, New York

Monge P, Lee S, Fulk J, Frank L, Margolin D, Schultz C, Shen C, Weber M (2011) Evolutionary and ecological models. In: Miller VD, Poole MS, Seibold DR, Myers KK, Park HS, Monge P, Fulk J, Frank LB, Margolin DB, Schultz CM, Shen C, Weber M, Lee S, Shumate M, Advancing research in organizational communication through quantitative methodology. Manage Commun Q 25:4–58

Mowshowitz A (1994) Virtual organization: a vision of management in the information age. Inf Soc 10:267–288

Mueller F, Harvey C, Howorth C (2003) The contestation of archetypes: negotiating scripts in a UK hospital trust board. J Manag Stud 40(8):1971–1995

Nair A, Filer L (2003) Cointegration of firm strategies within groups: a long-run analysis of firm behavior in the Japanese steel industry. Strateg Manag J 24:145–159

Payne GT (2006) Examining configurations and firm performance in a suboptimal equifinality context. Organ Sci 17:756–770

Pegels CC, Song YI, Yang B (2000) Management heterogeneity, competitive interaction groups, and firm performance. Strateg Manag J 21(9):911–923

Peng MW, Tan J, Tong TW (2004) Ownership types and strategic groups in an emerging economy. J Manag Stud 41:1105–1129

Pullman ME, Dillard J (2010) Values based supply chain management and emergent organizational structures. Int J Oper Prod Manag 30:744–771

Rauch A, Wiklund J, Freese M, Lumpkin GT (2004) Entrepreneurial orientation and business performance: cumulative empirical evidence. In: Zahra SA, Brush CG, Davidsson P, Fiet J, Greene PG, Harrison RT, Lerner M, Meyer GD, Sohl J, Zacharakis A, Mason C (eds) Frontiers of entrepreneurship research 2004. Babson, Wellesley, pp 164–177

Robins G, Pattison P, Kalish Y, Lusher D (2007) An introduction to exponential random graph (p*) models for social networks. Soc Networks 29(2):173–191

Schonpflug W, Luer G (2013) Science in a communist country: the case of the XXIInd international congress of psychology in Leipzig. Hist Psychol 16(2):112–129

Shi B (2010) The comparative research on two emergent organizational structures based on structure entropy model. In: Van de Walle B, Chu Z, Hu Y (eds) Proceedings of the 4th international ISCRAM-CHINA conference. Harbin Engineering University, Harbin, p 2010

Short JC, Payne GT, Ketchen DJ (2008) Research on organizational configurations: past accomplishments and future challenges. J Manag 34:1053–1079

Shumate M, Palazzolo E (2010) Exponential random graph (p*) models as a method for social network analysis in communication research. Commun Methods Meas 4:341–371

Siggelkow N (2002) Evolution toward fit. Adm Sci Q 47:125–159

Škerlavaj M, Dimovski V (2013) Social network approach to organizational learning. J Appl Bus Res 22(2):89–98

Snijders TAB, Pattison PE, Robins GL, Handcock MS (2006) New specifications for exponential random graph models. Sociol Methodol 36:99–153

Spanos YE, Zaralis G, Lioukas S (2004) Strategy and industry effect on profitability: evidence from Greece. Strateg Manag J 25:139–165

Stabell CB, Fjeldstad ØD (1998) Configuring value for competitive advantage: on chains, shops, and networks. Strateg Manag J 19(5):413–437

Stordeur S, D'Hoore W (2007) Organizational configuration of hospitals succeeding in attracting and retaining nurses. J Adv Nurs 57(1):45–58

de Toni AF, Nonino F (2010) The key roles in the informal organization: a network analysis perspective. Learn Organ 17(1):86–103

Tyrrall D, Parker D (2005) The fragmentation of a railway: a study of organizational change. J Manag Stud 42:509–537

Uhl-Bien M (2006) Rational leadership theory: exploring the social processes of leadership and organizing. Leadersh Q 17:654–676

Van de Ven AH, Poole MS (1995) Explaining development and change in organizations. Acad Manag Rev 20:510–540

Veliyath R, Shortell SM (1993) Strategic orientation, strategic planning system characteristics and performance. J Manag Stud 30(3):359–381

Weber M (1946) From max weber: essays in sociology. Oxford University Press, New York

Wiklund J, Shepherd D (2005) Entrepreneurial orientation and small business performance: a configurational approach. J Bus Ventur 20:71–91

Winter SJ, Taylor SL (1996) The role of IT in the transformation of work: a comparison of post-industrial, industrial, and proto-industrial organizations. Inf Syst Res 7:5–21

Youndt MA, Snell SA (2004) Human resource configurations, intellectual capital, and organizational performance. J Manag Issues 3:337–360

Chapter 7
Organizational Configurations and Configurational Change

7.1 Introduction

The first step in setting up the synthesized theoretical framework described in Chap. 6, which lays the groundwork for analyzing and understanding the full process of emergent organizational change—especially the changes observed in online organizations like Wikipedia—is to unite the storied domain of organizational configurations with the basic change dichotomy of evolution and revolution. This initial interweaving of concepts will facilitate the subsequent integration of organizational change motors, thereby yielding a powerful theoretical framework to serve future work in this area.

As such, this chapter explores the area of organizational configurations and links it with the evolutionary and revolutionary changes that we may observe in those configurations. We start by assessing the model laid out by Mintzberg (1979), significantly expanding on it by developing a comprehensive protocol to align any given organization with Mintzberg's theoretical archetypes. This is then connected, on both a conceptual and a measurement level, with the evolutionary and revolutionary changes that we may observe in those configurations, thereby yielding a new approach for researchers to directly observe and comprehend the development of organizational configurations over time. This is a vital piece of our larger theoretical model that specifies the mechanisms through which social structuration occurs in online collaborative groups.

7.2 Organizational Configurations

7.2.1 Configurations and Archetypes

One of the most important yet controversial methods for studying organizations is classifying them based on their respective configurations. A considerable amount of research has been devoted to this topic over the past 30 years, with multiple

© Springer International Publishing AG 2017
S.A. Matei, B.C. Britt, *Structural Differentiation in Social Media*, Lecture Notes in Social Networks, DOI 10.1007/978-3-319-64425-7_7

taxonomies and typologies emerging (see McPhee and Poole 2001 for an extensive review). This domain represents an attempt by scholars to move past analyzing where an organization falls on the spectra of a limitless number of attributes and toward synthesizing theory into a finite set of organizational forms which may be readily used for descriptive and prescriptive purposes.

The main allure of such classification schemes is their parsimony. Condensing the results of countless organizational measures into a few critical archetypes allows practitioners to steer their organizations toward a desired, time-tested structure or at least to coax an emergent structure in a positive direction from above or from within. This is a much more effective strategy than merely working to match the scores of quantitative metrics which, in the first place, only indirectly tap into the underlying constructs of interest. Configurations, seen as concepts, are also useful for analysts trying to study differences between subsets of organizations, as those subsets can be constructed heuristically based on matching configurations, and those configurations themselves typically integrate many otherwise divergent bodies of literature on various aspects of organizational structure.

Still, this approach has its share of detractors (see McPhee and Poole 2001, 514–515). In particular, as McPhee and Poole note, many scholars vigorously challenge the notion of casting organizational structures into rigid groups, rightly arguing that no real organization can be expected to represent a pure version of any classification. Given this fact, we sacrifice a great deal of information when we ignore the impurities in any organization's true configuration by defining it based upon an archetype.

Opponents of such classification systems therefore advocate treating the factors that would define a typology as spectra rather than exclusive groups. As they argue, just as restricting the concept of "color" to the seven well-known shades of the rainbow would needlessly cast aside all the other tints and hues, arbitrarily limiting the literature to a finite number of organizational structures would necessarily mean discarding an infinite number of intermediate and combination forms.

It is possible, however, to achieve the benefits of existing organizational classifications while avoiding the trap of overclassification. Rather than merely seeking the configurational archetype closest to a given organization and defining it accordingly, we may instead examine several commonly measured organizational characteristics and then use them to describe the configuration, or variation, that the organization holds. Because these measures themselves take on spectra of possible values, they may be used to discern the degree to which an organization resembles a range of different configurations rather than wrongly treating it as a perfect representation of a single archetype. In other words, the pure configurations are treated as poles on a spectrum, and this approach simply measures how closely the organization in question fits each of those poles. To use our earlier analogy, while we are certainly right to acknowledge that there are infinite shades between orange and red, it is still useful to have something that we can just call "red."

Additionally, using common organizational measures to classify organizations into configurations provides a secondary benefit in its own right, greatly simplifying the often-complicated classification process. The approach articulated in this chapter may therefore serve as a guide for others who want a parsimonious, straightforward method of classifying organizations of all types.

7.2.2 Mintzberg's Configurational Array

Henry Mintzberg (1979, 1989) is responsible for what is arguably the most highly respected and widely used set of organizational configurations in the literature (for prominent examples of use, see Boumgarden et al. 2012; Finkelstein et al. 2009; Fiss 2011; Ireland et al. 2009; McPhee and Zaug 2009; Rainey 2003). Mintzberg's Configurational Array, as it has come to be known, shares many similarities with prior taxonomies of organizational configurations, combining them into a theoretically satisfying system of organizational sectors, processes, and configurations that formed the foundation for much of the modern configuration literature.

Mintzberg's typology focuses on the structural form that an organization holds, as well as the processes and systems that support and maintain that structure, allowing it to be more easily generalized across domains than other configurational typologies that focus more on the size of the organization, the industry in which it is embedded, and other context-specific features. This chapter refers to Mintzberg's configurations as "organizational configurations" or simply "configurations" in order to maintain consistency with the literature. However, to be clear, Mintzberg's organizational configurations are defined by the organization's structural form, so these structural forms represent the focal point of much of the present chapter.

With that in mind, Mintzberg's Configurational Array was originally (1979) grounded in five organizational coordination mechanisms: direct supervision, standardization of work processes, standardization of work outputs, standardization of work skills, and mutual adjustment through informal communication. These formed the foundation for his five original configurational archetypes, which he termed simple structure (or entrepreneurial), machine bureaucracy, professional bureaucracy, diversified (or divisionalized), and adhocracy, the last of which, as we have previously mentioned, takes on a much more specific definition here than the generic concept used in Part I of this book.

Briefly, the entrepreneurial organization resembles the classic "mom-and-pop" store, with a single leader directly managing a collection of loosely organized subordinates. This stands in contrast with the more rigid, hierarchical machine bureaucracy, as well as the professional bureaucracy, the latter of which relies more on the efforts of its lower-level members, the "operating core," than the hierarchy through which orders are disseminated from the top to the bottom. The diversified organization is split into a number of nearly autonomous divisions, operating much like franchises of a restaurant chain, thus drawing control away from the top toward each individual unit. Finally, an adhocracy features project teams that overlap with one another and may disperse and reform organically, without clearly defined, permanently occupied roles. Notably, structure is not absent from adhocracies; what makes them different is the ad hoc nature of their structures and the order in which various actors occupy an assortment of positions and roles over time.

Mintzberg later (1989) added two additional processes—the standardization of norms and political conflict (the latter of which opposes coordination)—which are essential for two further configurations, the missionary organization and the political organization, respectively. However, the political configuration is rarely seen outside of the political realm itself, as most organizations would quickly disintegrate if their dominant processes were those of political warfare. For that matter, political systems are generally comprised of multiple parties with mutually exclusive goals, in contrast with most organizations that have at least one primary shared objective that pervades the system. Since calling a political system an "organization" would strip the term of much of its meaning, most scholars have since come to ignore the political configuration.

Similarly, a missionary organization generally has its ineffable mission instilled within it from the beginning such that only a very specific set of organizations, such as religious sects, can ever be realistically classified as missionary organizations. As such, much like the political configuration, many scholars ignore the missionary configuration due to similar limitations and focus only on the five core configurations that Mintzberg (1979) originally devised.

As Mintzberg argued, different parts of an organization may use various coordination mechanisms to exert pressure on the organization itself and pull it toward a particular manner of functioning. For instance, the chain of middle management, or "middle line," as Mintzberg called it, may take advantage of its ability to directly supervise other workers, with each manager systematically issuing orders to a group of subordinates, in order to push the organization toward the machine bureaucracy archetype. It should come as no surprise, then, that any given part of an organization tends to be associated with a single coordination mechanism, which in turn drives the organization toward one particular configuration.

More broadly, the five primary archetypes and the processes that generate them are directly related to several structural dynamics that occur within organizations (Mintzberg 1979). The leader who stands atop a simple structure, for instance, holds command over decision-making and stands at the peak of a free-form, organic structure with minimal differentiation among the other organizational members. The machine bureaucracy maintains that locus of power at the top of the organization, but with its leaders imposing a fixed hierarchy upon subordinates such that there is minimal intermixing among unrelated work groups. On the other hand, the professional bureaucracy is characterized by a great deal of control being ceded to the operating core at the bottom of the hierarchy, so organizational power is not nearly as localized among a small clique of leaders and behavior is not as formalized as it is within the more rigid machine hierarchy.

Within a diversified organization, the structure is fragmented into a number of largely distinct divisions, each of which maintains its own structure and strict standards for activity such that the organization's activities as a whole are somewhat decentralized, yet the leaders continue to serve as a vital link between the disparate units. And lastly, the final archetype, that of an adhocracy, features an organic structure similar to that of the simple structure, but with the flexibility to form self-directed project teams as needed based on the specialized skills of their members and without an entrenched leader necessarily dominating the structure from the top. Yet, adhocracy should not be confused with a total lack of organizational structure. Again, its distinguishing feature is the flexibility of roles and attributes, which is supported by the members' ability and willingness to take over specific roles and abandon them as needed.

Importantly, Mintzberg's Configurational Array is defined by the extent to which a particular component pulls the organization toward a structure that benefits itself, which implicitly allows the typology to handle hybrid or in-between organizational configurations well—these, after all, would be represented by the transition or struggle for control between competing sectors. As such, Mintzberg's typology avoids falling into the trap of needlessly forcing all organizations to fit within a finite set of overly rigid, extreme archetypes, instead allowing for a spectrum of organizations to be observed. Furthermore, regardless of the organizational structure that is formally prescribed, any functional configuration might ultimately emerge as the dominant mechanism, as different parts of the organization may pull the system into a configuration that better serves their respective designs.

In short, Mintzberg's Configurational Array is perhaps the single most prominent set of organizational configurations in the literature, offering a useful, theoretically driven synthesis of many other classification systems. It offers researchers a straightforward way to conceptualize and identify various organizational configurations, and it grants practitioners a natural mechanism for evaluating and managing their own organizations. The typology makes use of several well-defined and often-implemented organizational forms, and by focusing on the influences that shape a configuration rather than merely the end result, it effectively accounts for hybrid organizational forms and facilitates the observation of a range of intermediate classifications between the archetypes. As such, Mintzberg's Configurational Array represents an optimal tool to observe and define an organization's configuration as well as its transitions between distinct configurations.

7.3 Configurational Change

7.3.1 Configurations and Change in Virtual Organizations

Our next major step, then, is to draw a conceptual connection between organizational configurations and organizational change. Yet in order to understand the importance of combining these two frameworks, one must first recognize the limitations of classifying organizations using static configurations alone.

Online groups such as Wikipedia serve as a particularly useful example of such limitations, as one of the biggest disputes among researchers who have studied Wikipedia is the particular configuration that it holds. Much of this debate originally stemmed from Wikipedia's longtime slogan, which labeled it "The free online encyclopedia that anyone can edit" (Fallis 2008; Wikipedia 2012) and which implied that such an open organization would eschew any restrictive organizational hierarchies and result in equal rights among all members. Some (e.g., Faucher et al. 2008; Jemielniak 2012) have argued that Wikipedia's lack of formal stratification in its organizational structure and the prioritization of consensus over control necessarily imply a lack of order, with members engaging with one another within an anarchy that is largely free of rulings imposed from above. Besides, as Mateos-Garcia and Steinmueller (2006) pointed out, a number of top contributors have perceived at least a degree of anarchism within Wikipedia, as has founder Larry Sanger (2006, 323; see also Wikimedia 2011), further evoking notions of egalitarianism and the so-called wisdom of crowds (Shirky 2008; Surowiecki 2004).

Others have belied this notion and claimed that, contrary to popular belief, the system of Wikipedia editors closely resembles a formalized bureaucracy, in which a few key individuals hold all the power and command those at the lower levels of the organization, just as we are accustomed to observing and experiencing in traditional hierarchies. Such scholars, such as Butler et al. (2008), have noted that Wikipedia holds a relatively small administrative staff (1,281 administrators, 22 bureaucrats, and 30 stewards as of January 29, 2017), a small subset of the millions of Wikipedia contributors. These leaders collectively designed the community's rules and arbitrate any disputes, taking the authority out of the hands of the masses (Wikimedia 2013; Wikipedia 2013a, b). Viégas et al. (2007) likewise argued that the wiki framework itself is uniquely suited to authoritative control. Beyond this, a number of researchers have noted the prevalence of the Pareto principle on Wikipedia (e.g., Bruno 2010; Ehmann et al. 2008; He 2011; Matei and Bruno 2015), ultimately demonstrating that 80–90% of the encyclopedic content was generated by a minority group encompassing 1% of the editors (see Ortega et al. 2008 for the early stages and Matei and Bruno 2015 for more recent history). Some may argue that these findings suggest that the organization's power is localized at the top of a relatively strict hierarchy, even if that hierarchy rests beneath the surface of the organization (Matei and Dobrescu 2011). Finally, to further drive home the point, Müller-Seitz and Reger (2010) showed that even Wikipedia contributors themselves perceive their organization to be bureaucratic, in stark contrast with the flexible anarchic system that others have long suspected the community represents.

But this perception, too, is contested. A few scholars (e.g., Jameson 2011; Konieczny 2010) have more recently claimed that Wikipedia instead represents an adhocracy thanks to its organic structure and lack of formalization (see Mintzberg 1989). While Wikipedia indeed has a structure, these scholars argue, it is not explicitly defined in the same manner as a bureaucracy, so it may freely change to suit its own needs (Matei and Dobrescu 2011). Furthermore, the division of Wikipedia into millions of articles, or focused "project teams," would also seem to align with the notion of an adhocratic structure (Pamkowska 2008). It is possible that an elite group eventually solidified at the top of the organization, as Ortega et al. (2008) argued, or that the top contributors change from week to week, a belief that many who believe in an anarchic Wikipedia espouse. Scholars like Konieczny (2010) and Matei and Dobrescu (2011), however, claim that any such leadership roles are dynamic, emerging from behavior within the organization and subject to significant change over time, rather than representing inscribed positions defined by a set of bureaucratic heads.

Clearly, Wikipedia appears to share traits with many different organizational forms. Even Wikipedians themselves have been unable to conclusively determine what configuration their own system follows (Wikimedia 2012), especially since numerous alternative terms such as republic, despotism, and plutocracy have been proposed to describe the community. It is little wonder, then, that scholars have long debated which theoretical form is the most accurate representation of the knowledge-construction community (see O'Neil 2010 for a focused review of the debate). In light of Wikipedia's importance to users around the world, it would certainly behoove organizational scholars to conclusively determine this fundamental detail about how the system as a whole operates.

Yet in truth, the bigger point is not merely the question of whether Wikipedia is an anarchy, a bureaucracy, an adhocracy, or something else. It isn't even that scholars have been unable to agree on what, exactly, Wikipedia and other online collaboration groups are. Rather, the point is that all of these arguments over online organizational configurations miss a critical issue: online collaborative groups, like any other organization, are not static. They have grown considerably over the years, and much like the McDonald's fast-food chain now hardly resembles the single restaurant from which the international powerhouse originated (Mieth 2007), it is likely that the structures of online collaborative groups, especially Wikipedia, have undergone substantial transformations over the course of their development.

As such, we are free to speculate that Wikipedia, as well as any other collaborative group, may have been an anarchy, a bureaucracy, and an adhocracy at various points in time, and it may have also held any number of other organizational configurations that one might envision. Furthermore, it could be that Wikipedia was founded in such a way that it evolved along a substantially different course than other organizations that have failed, moving toward a more novel organizational form than its fallen peers, or it could be that Wikipedia, like many traditional offline organizations, became increasingly rigid and bureaucratic as it expanded, and that it simply emulated particular aspects of an archetypal organizational form better than its competitors.

In order to clarify the organizational configuration or configurations that online collaborative groups may have held over time, we will consider the case of Wikipedia on a much deeper level. Understanding the transitions that an organization makes between different organizational configurations, after all, necessarily illuminates the organizational configurations themselves. Further, prior studies of Wikipedia's early years (Britt 2011; Matei et al. 2015) suggest that the structure of Wikipedia progressed through transitional stages over the course of its development, so it is reasonable to suspect that the Wikipedia coeditorial community may have also cycled through a number of different configurations over time. This would, after all, explain why there is such little agreement about the particular organizational configuration that Wikipedia, which is often envisioned as a static, monolithic entity, holds. It is with this in mind that we begin to build upon the framework of organizational configurations by turning to the two primary classes of organizational change: evolution and revolution.

7.3.2   Evolutionary and Revolutionary Change

Many organizational change scholars (e.g., Greiner 1972, 1994; Mezias and Glynn 1993) have historically dichotomized the concept of change into two primary classes: evolution and revolution. The distinction in the dichotomy is largely a matter of timing, or the speed at which significant change occurs, whether modest and gradual or dramatic and immediate (Gersick 1991; Levy and Merry 1986).

Revolution, as the term implies, denotes moments of rapid, significant change, akin to the act of deposing a king or dictator, which can sharply divert the course of a nation's history. As Greiner (1972) defines it, "The term *revolution* is used to describe those periods of substantial turmoil in organization life." Relative to the long-term growth that we might otherwise observe, such a revolution generally occurs in an instant, as a sudden, decisive moment in history. As a result of the revolution, be it the overhauling of a corporate hierarchy or the overthrowing of a ruler, the organization is forever changed. This change may constitute the reassignment or reversal of roles for members of the organization, or it could entail the development of a completely new role or system of roles altogether, as evidenced by changes in some central measures of social organization and structuration, such as entropy, as suggested above.

This stands in contrast with much slower evolutionary changes that occur gradually, as the organization shifts slowly, almost imperceptibly, toward a new standard, norm, policy, framework, or goal. For instance, daily tasks may become entrenched as regimented routines, accruing value and subtly pulling the organization toward actions that would maintain them as a standard practice, or they may instead falter over time as organizational members find merit in alternative approaches and incrementally deviate from the original routine, with those members pushing the organization away from past methods and toward standards that they deem superior. While revolution implies a sharp change in trajectory, we may conceive of evolution as an organization following its existing momentum for a period of time (Miller and Friesen 1980), with greater stability in behavioral norms and in the operational characteristics of the organization in the long term.

Revolutionary and evolutionary change offer, as a pair, a comprehensive representation of all organizational change. Either an organization changes instantly, with the community suddenly shocked into a new mode or form, or it undergoes a much smoother, incremental change. The difference is not strictly one of time but of the degree of structural change—in other words, the contrast is that between an abrupt break and a gradual curve, one of which is a much more severe, jarring shift than the other. Some evaluation may be necessary to properly classify observed changes into these two groups, of course, but the simplicity of this dichotomy, with the categories of change very broadly defined, is what allows for its comprehensiveness.

As a caveat, some scholars (especially Levy and Merry 1986) argue for a continuum of change between evolution and revolution rather than the present dichotomy of timings. However, these arguments have largely been grounded in the dual ideas that evolutionary change may be too slow for managers to deem it a successful strategy and that revolutions can likewise damage an organization, rather than substantial evidence that the evolutionary-revolutionary dichotomy is incorrect. In other words, the arguments against this dichotomy are based only on how some scholars think organizations ought to operate as opposed to what researchers have actually observed.

More importantly, organizations may very well be subject to both types of change, with gradual evolutionary trends punctuated by moments of revolution. These two types of change, then, combine to represent the full progression of organizational change as it occurs over time. Greiner (1972) offered one of the most famous examples of such a model, presenting a growth process consisting of five evolutionary phases culminating in crises that must be overcome through a suitable revolution. With this conceptualization in mind, one could presumably measure any particular organizational attribute over time, chart the way it changed during the organization's development, and quickly identify those periods of evolutionary change (stable linear and curvilinear trends) and revolutionary change (sharp jumps or breaks from the prior trend). This conceptualization is used below to describe Wikipedia and, by extension, similar online collaborative communities.

7.3.3 Pinpointing Configurational Evolution and Revolution

In what follows we will propose specific conceptual perspectives and measurements to characterize organizational configurations. First and foremost, the defining characteristics of organizational configurations can be assessed using five key organizational dimensions: the structures of collaborative attractiveness and extroversion, communication flow, partner choice trends, and structural order.

Collaborative attractiveness is the extent to which organizational members are drawn toward interactions with a few particular elites, while collaborative extroversion refers to the efforts of a few especially active individuals to seek out interactions with others throughout the community, well beyond the interaction-seeking behaviors of their peers. Communication flow describes the extent to which a small set of especially well-positioned members control the interactions between larger constituencies by serving as the primary conduits connecting otherwise largely distinct groups. The partner choice trends dimension refers to the tendency for organizational members to form and build relationships with specific types of peers based on mutual similarities or differences. Finally, structural order describes the extent to which an elite group dominates the collaborative process such that contributions and the resulting content are centered on their efforts, above and beyond those of other contributors.

These dimensions jointly reflect the five coordination mechanisms of direct supervision, standardization of work processes, standardization of work outputs, standardization of work skills, and mutual adjustment through informal communication that define Mintzberg's (1979) configurational archetypes.

Briefly, direct supervision, the cornerstone of a simple structure, allows a subset of leaders to directly oversee subordinates, becoming the most vocal (extroversion) members of the organization as well as those who receive the most communicative feedback (attractiveness) and allowing them to assign and reassign tasks and teams on a regular basis (structural order). However, given that other organizational members would still need to interact with one another in order to perform their roles, direct supervision would not necessarily constrain communication flow, especially if teams were indeed shuffled over time. Likewise, since individual members would have little control over the peers with whom they would interact at any given moment, no partner choice trends would be likely to have significant relevance within such a model.

In contrast, the standardization of processes, which stands as the hallmark of the machine bureaucracy archetype, may only be carried out when a set of leaders takes command of information dissemination within the organization (communication flow) to direct activities accordingly. Such leaders are likely to find themselves as the most frequent communicators (extroversion) as well as those who receive the most information from within the organization (attractiveness), though this would likely be more of an unintended consequence of controlling the communication flow rather than the leaders' intention. On the other hand, standardizing work processes would demand a high degree of structural order, and as work processes became increasingly standardized, those performing similar roles would likely be grouped together (partner choice trends) in order to further standardize their processes.

The standardization of outputs, which pushes an organization toward a divisionalized form, still demands the separation of members performing different tasks or otherwise operating in different branches of the organization (partner choice trends), and as those distinct units come to engage in minimal direct interaction with one another, the organizational leaders assume even greater control over information dissemination (communication flow) than in a machine bureaucracy. Such a configuration must be highly structured, of course (structural order), yet as the various units gain increasing autonomy from the rest of the organization, the leaders play a declining role in directing the day-to-day activities of other members (extroversion) and receiving feedback thereafter (attractiveness).

When skills are standardized, on the other hand, there is little need for an extensive leadership structure that would direct (extroversion) and track (attractiveness) the organizational activities, nor would such leaders have any need to guide the dissemination of information throughout the organization (communication flow). Yet, as such a mechanism would lend itself to a professional bureaucracy, the form of the organization would still be well defined (structural order), and those members who perform similar roles or have similar functional specialties (see Mintzberg 1979, 354) would likely still be grouped together (partner choice trends).

Finally, the act of mutual adjustment through informal communication necessarily implies a lack of consistent structural order. Like the divisionalized organization, there would be no substantial need for leaders to directly manage the process, meaning that structures of attractiveness and extroversion would hold minimal importance. Further, since members would coordinate themselves rather than relying on an entrenched leadership cadre to direct their activities, any leaders that were present would hold minimal control over information distribution across the organization (communication flow). Similarly, since members would frequently adjust their associations with others, rapidly rotating between different small groups in an adhocratic manner, we would not expect members to consistently associate with similar peers; in fact, members who tend to fill gaps in functional roles within various groups might consequently be more likely to interact with peers who are substantially different than themselves (partner choice trends).

As illustrated here, the five dimensions of collaborative attractiveness and extroversion structures, communication flow, partner choice trends, and structural order jointly reflect the key processes that define Mintzberg's (1979) five configurational archetypes by further decomposing the associated coordinating mechanisms into their behavioral components.

These five key constructs, in turn, are themselves associated with five simple measurements: inbound degree centralization, outbound degree centralization, betweenness centralization, assortativity, and entropy. Inbound and outbound degree centralization, as defined by Freeman (1979), offer a direct summary of how much communication is localized around a few key individuals. Both of these measures assess how much the communicative activity within an organization is localized around a few key individuals, and their measurements effectively serve as ratios between the communicative activity of the most prominent organizational members and the activity of the entire system as a whole. In short, inbound and outbound degree centralization serve as operationalizations of the structures of attractiveness and extroversion, respectively, within a given organization.

While high outbound degree centralization would indicate that a small set of individuals speak directly to most of the organization, whether to informally converse or to issue explicit directives, high inbound degree centralization would suggest that the "masses" of the organization seek out a few prominent individuals and that those leaders are the targets of much more communication than the rest of the organization. In a hierarchy, this would be the difference between top-down and bottom-up communication. Depending on other structural factors, declining inbound and outbound degree centralization could suggest a more extensive hierarchy, with more messages between the operating core and the leadership apex proceeding through or being filtered by intermediate layers of the organization, or it could instead indicate the erosion of the leadership position as the organization flattens and minimizes its reliance on authority from above.

One of those other structural factors is betweenness centralization, which, like the two forms of degree centralization, indicates how much the organization's power is focused around a few key individuals. Unlike degree centralization, however, betweenness centralization describes the extent to which the flow of communication through the organization must go through particular individuals and, therefore, how much control the most prominent individuals have over information in contrast with all the other organizational members. In a rigid hierarchy with wholly distinct divisions, for instance, the individual coordinating those divisions from the top would wield tremendous power over the flow of communication through the organization, as those distinct divisions would have no means to communicate with one another except through their leader. The more rigid the organizational hierarchy grows, the more that communication between disparate units must travel through the apex of that hierarchy, and the greater the organization's betweenness centralization will be.

It is also important to assess the degree to which organizational members are restricted to interacting with peers in their particular subgroup, or, in other words, how much intermixing occurs across organizational levels (i.e., the partner choice trends within the organization). The extent to which individuals interact with those who are similar to them, as opposed to those who hold a very different standing in the organization, may be evaluated using assortativity (Newman 2002). When organizational leaders tend to interact primarily with fellow leaders and their subordinates, likewise, are drawn to interact with fellow subordinates, the organizational network is considered assortative. If, instead, the organization's leaders engage more with those at the bottom of the organization and vice versa, the network is more disassortative, with its members routinely crossing traditional status boundaries that might otherwise prevent, for instance, a factory worker from interacting extensively with his company's CEO.

Lastly, the concept of order within the community structure can be operationalized using social entropy which, as described by Matei et al. (2010a) and expanded upon in Chap. 4, indicates the balance of contributions made to the system (see also Bailey 1990; Britt 2013; Shannon and Weaver 1948). When a small subset of the population takes control of the collaborative effort and contributes most of the product on its own, the organization becomes highly structured around this core group of individuals. If, on the other hand, contributions are made relatively evenly across the population, no individual or group clearly rises to power, making the organizational structure flatter and the power dynamic much more entropic. So as a general rule, the more entropic the community, the less structured it is and the less predictable future activity is.

While equality in contributions across group members is often colloquially considered to represent pure, perfect collaboration (see Chap. 4), numerous studies have shown that in most groups a small core of individuals dominates the interactions and, further, that a degree of inequality in contributions actually improves collaboration in both its process and its product (e.g., Bruno 2010; Kittur et al. 2007; Kuk 2006; Shirky 2008). Although a highly structured organization may waste excessive energy maintaining its structure, a completely free-form or flat organization can find its efforts stymied by the lack of any clear leadership or direction as well as redundancy, so some balance between these two structural extremes has been proposed as an ideal state for collaboration (Bruno 2010; Kittur et al. 2007; Matei et al. 2010b; Shirky 2008). Notably, a degree of contribution inequity has previously been found on Wikipedia, with a small set of leading contributors driving the majority of the knowledge-construction work (Bruno 2010; Ehmann et al. 2008; He 2011; Matei et al. 2015; Ortega et al. 2008), but the stability of this inequity remains unclear.

Together, these five constructs allow us to directly observe the extent to which a given organization follows each of Mintzberg's (1989) configurational archetypes.

	Inbound Degree Centralization	Outbound Degree Centralization	Betweenness Centralization	Assortativity	Entropy
	Structure of Collaborative Attractiveness	Structure of Collaborative Extroversion	Structure of Communication Flows	Partner Choice Trends	Structural Order
Simple Structure Entrepreneurial	Very High	Very High	Low	Zero/Negative	Somewhat High
Machine Bureaucracy	Somewhat High	Somewhat High	High	Positive	Low
Professional Bureaucracy	Low	Low	Low	Positive	Low
Diversified Divisionalized	Somewhat Low	Somewhat Low	Very High	Positive	Low
Adhocracy	Low	Low	Low	Zero	Moderate

Fig. 7.1 Measurement levels for Mintzberg's configurational archetypes

More specifically, the five organizational dimensions and their accompanying metrics can be used as a multidimensional scale to characterize these configurations.

Figure 7.1 illustrates the combinations of organizational levels that best represent each of the five archetypes. Exact numerical figures are not given for measurements of these constructs in order to facilitate easier interpretations, but we may generally say that the more closely that a given organization adheres to a particular combination, the purer a representation it is of the associated configuration.

It may be noted that we would expect the observed patterns of inbound degree centralization and outbound degree centralization to be very similar to one another within any of the configurational archetypes. Nonetheless, past research (e.g., Britt et al. 2011) suggests that it may be worthwhile to examine inbound degree centralization and outbound degree centralization separately rather than to combine them into a single degree centralization measure. In particular, one form of degree centralization might shift before the other, offering an earlier indication of a configurational change than what may be observed through a combined assessment, as a unified degree centralization measure may mask changes to either of its individual components. Furthermore, while inbound and outbound degree centralization should be virtually identical within any of Mintzberg's archetypes, the same may not necessarily be true of configurations lying on the spectrum between those archetypes, so using both measures facilitates the characterization of those intermediate organizational forms.

7.4 Conclusion

As a whole, this chapter serves as a theoretical and methodological unification of Mintzberg's Configurational Array (1979) with the conceptual pair of evolution and revolution, using five common organizational measures as a tool for classification purposes. On its surface, this extension of the configuration framework is relatively simple, as evolution and revolution merely describe the ways in which an organization moves from one configuration to the next. Yet, this seemingly basic conceptual integration is essential to enable the change dichotomy to serve as a conduit that ultimately connects organizational configurations with organizational change motors. With that in mind, the incorporation of organizational change motors, which completes the unified theoretical framework, is addressed at length in the next chapter.

References

Bailey KD (1990) Social entropy theory. State University of New York Press, Albany

Boumgarden P, Nickerson J, Zenger TR (2012) Sailing into the wind: exploring the relationships among ambidexterity, vacillation, and organizational performance. Strateg Manag J 33:587–610

Britt BC (2011) System-level motivating factors for collaboration on Wikipedia: a longitudinal network analysis. Thesis, Purdue University

Britt BC (2013) Evolution and revolution of organizational configurations on Wikipedia: a longitudinal network analysis. Dissertation, Purdue University

Britt BC, Matei SA, Braun D (2011) Mining large-scale online communities: the development and dispersion of tools for analyzing collaborative processes and structures. Paper presented at the Sunbelt Social Network Conference XXXI, St. Pete Beach, 8–13 Feb 2011

Bruno RJ (2010) Social differentiation, participation inequality and optimal collaborative learning online. Dissertation, Purdue University

Butler B, Joyce E, Pike J (2008) Don't look now, but we've created a bureaucracy: the nature and roles of policies and rules in Wikipedia. In: Czerwinski M, Lund A (eds) Proceedings of the SIGCHI conference on human factors in computing systems (CHI 2008). ACM Press, New York

Ehmann K, Large A, Behesti J (2008) Collaboration in context: comparing article evolution among subject disciplines in Wikipedia. First Monday 13(10). https://doi.org/10.5210/fm.v13i10.2217. http://firstmonday.org/article/view/2217/2034

Fallis D (2008) Toward an epistemology of Wikipedia. J Am Soc Inf Sci Technol 59(10):1662–1674

Faucher J-BPL, Everett AM, Lawson R (2008) A complex adaptive organization under the lens of the LIFE model: the case of Wikipedia. In: Gray B (ed) Proceedings of the 11th McGill international entrepreneurship conference, Dunedin, 2008

Finkelstein S, Hambrick DC, Cannella AA Jr (2009) Strategic leadership: theory and research on executive, top management teams, and boards. Oxford University Press, New York

Fiss PC (2011) Building better causal theories: a fuzzy set approach to typologies in organization research. Acad Manag J 54:393–420

Freeman LC (1979) Centrality in networks: I. Conceptual clarification. Soc Networks 1:215–239

Gersick CJ (1991) Revolutionary change theories: a multilevel exploration of the punctuated equilibrium paradigm. Acad Manag Rev 16:10–36

Greiner LE (1972) Evolution and revolution as organizations grow. Harv Bus Rev 50(4):37–46

Greiner LE (1994) Evolution and revolution as organizations grow. In: Mainiero L, Tromley C (eds) Developing managerial skills in organizational behavior: exercises, cases, and readings, 2nd edn. Prentice Hall, Englewood Cliffs, pp 322–329

He Z (2011) Measuring the development of Wikipedia. Paper presented at the 2nd international conference on Internet technology and applications, Wuhan, 16–18 Aug 2011

Ireland RD, Covin JG, Kuratko DF (2009) Conceptualizing corporate entrepreneurship strategy. Entrep Theory Pract 33(1):19–46

Jameson J (2011) Leadership of shared spaces in online learning communities. Int J Web Based Communities 7:463–477

Jemielniak D (2012) Wikipedia: an effective anarchy. Paper presented at the society for applied anthropology 2012 annual meeting, Baltimore, 27–31 Mar 2012

Kittur A, Chi EH, Pendleton BA, Suh B, Mytkowicz T (2007) Power of the few vs. wisdom of the crowd: Wikipedia and the rise of the bourgeoisie. Paper presented at the 25th annual ACM conference on human factors in computing systems (CHI 2007), San Jose, 28 Apr-3 May 2007

Konieczny P (2010) Adhocratic governance in the internet age: a case of Wikipedia. J Inf Technol Polit 7(4):263–283

Kuk G (2006) Strategic interaction and knowledge sharing in the KDE developer mailing list. Manag Sci 52(7):1031–1042

Levy A, Merry U (1986) Organizational transformation: approaches, strategies, theories. Praeger Publishers, New York

Matei SA, Bruno RJ (2015) Pareto's 80/20 law and social differentiation: a social entropy perspective. Public Relat Rev 41(2):178–186

Matei SA, Dobrescu C (2011) Wikipedia's "neutral point of view": settling conflict through ambiguity. Inf Soc 27(1):40–51

Matei SA, Bruno RJ, Morris P (2010a) Visible effort: a social entropy methodology for managing computer-mediated collaborative learning. Paper presented at the Global Communication Forum, Shanghai, 29–30 Sept 2010

Matei SA, Oh K, Bruno R (2010b) Collaboration and communication in online environments: a social entropy approach. In: Oancea M (ed) Comunicare și comportament organizational (communication and organizational behavior). Printech, Bucharest, pp 82–98

Matei SA, Bertino E, Zhu M, Liu C, Si L, Britt BC (2015) A research agenda for the study of entropic social structural evolution, functional roles, adhocratic leadership styles, and credibility in online organizations and knowledge markets. In: Bertino E, Matei SA (eds) Roles, trust, and reputation in social media knowledge markets: theory and methods. Springer, New York, pp 3–33

Mateos-Garcia J, Steinmueller WE (2006) Open, but how much? Growth, conflict, and institutional evolution in Wikipedia and Debian. Paper presented at EU-DIME International Conference 2006, Durham, 27–28 Oct 2006

McPhee RD, Poole MS (2001) Organizational structures and configurations. In: Jablin FM, Putnam LL (eds) The new handbook of organizational communication: advances in theory, research, and methods. Sage, Thousand Oaks, pp 503–543

McPhee RD, Zaug P (2009) The communicative constitution of organizations. In: Putnam LL, Nicotera AM (eds) Building theories of organization: the constitutive role of communication. Routledge, New York, pp 21–48

Mezias SJ, Glynn MA (1993) The three faces of corporate renewal: institution, revolution, and evolution. Strateg Manag J 14:77–101

Mieth H (2007) The history of McDonald's. GRIN Publishing, Munich

Miller D, Friesen P (1980) Archetypes of organizational transition. Adm Sci Q 25:268–299

Mintzberg H (1979) The structuring of organizations: a synthesis of the research. Prentice Hall, Englewood Cliffs

Mintzberg H (1989) Mintzberg on management: inside our strategic world of organizations. Free Press, New York

Müller-Seitz G, Reger G (2010) 'Wikipedia, the free encyclopedia' as a role model? Lessons for open innovation from an exploratory examination of the supposedly democratic-anarchic nature of Wikipedia. Int J Technol Manag 52:457–476

Newman MEJ (2002) Assortative mixing in networks. Phys Rev Lett 89:1. article 208701

O'Neil M (2010) Wikipedia and authority. In: Lovink G, Tkacz N (eds) Critical point of view: a Wikipedia reader. Institute of Network Cultures, Amsterdam, pp 309–324

Ortega F, Gonzalez-Barahona JM, Robles G (2008) On the inequality of contributions to Wikipedia. In: Proceedings of the 41st Hawaii international conference on system sciences (HICSS '08), IEEE, Washington DC, 7–10 Jan 2008

Pamkowska M (2008) Autopoiesis in virtual organizations. Informatica Economică 1(45):33–39

Rainey HG (2003) Understanding & managing public organizations, 3rd edn. Jossey-Bass, San Francisco

Sanger L (2006) The early history of Nupedia and Wikipedia: a memoir. In: DiBona C, Stone M, Cooper D (eds) Open sources 2.0: the continuing evolution. O'Reilly Media, Sebastopol, pp 307–338

Shannon CE, Weaver W (1948) The mathematical theory of communication. University of Illinois Press, Urbana

Shirky C (2008) Here comes everybody: the power of organizing without organizations. Penguin, New York

Surowiecki J (2004) The wisdom of crowds: why the many are smarter than the few and how collective wisdom shapes business, economies, societies, and nations, 1st edn. Doubleday, New York

Viégas FB, Wattenberg M, McKeon MM (2007) The hidden order of Wikipedia. In: Schuler D (ed) Proceedings of the 2nd international conference on online communities and social computing. Springer, Heidelberg

Wikimedia (2011) Is Wikipedia an experiment in anarchy? http://meta.wikimedia.org/wiki/Is_Wikipedia_an_experiment_in_anarchy. Accessed 31 Jan 2017

Wikimedia (2012) Wikipedia power structure. http://meta.wikimedia.org/wiki/Power_structure. Accessed 4 Mar 2013

Wikimedia (2013) Stewards. http://meta.wikimedia.org/wiki/Steward. Accessed 29 Jan 2017

Wikipedia (2012) Wikipedia:The Free Encyclopedia. http://en.wikipedia.org/wiki/Wikipedia:The_Free_Encyclopedia. Accessed 8 Oct 2012

Wikipedia (2013a) Wikipedia:Bureaucrats. http://en.wikipedia.org/wiki/Wikipedia:Bureaucrats. Accessed 29 Jan 2017

Wikipedia (2013b) Wikipedia:List of administrators. http://en.wikipedia.org/wiki/Wikipedia:List_of_administrators. Accessed 29 Jan 2017

Chapter 8
A Synthesized Theoretical Framework for Motors Driving Organizational Configurational Change

8.1 Introduction

Now that we have a framework for understanding organizational configurations and the manner in which they change over time, we need to unite this with the forces that drive those configurational changes. To that end, this chapter introduces the concept of organizational change motors, as articulated by Van de Ven and Poole (1995), and then serves to synthesize the two historically distinct domains of organizational configurations and organizational change motors into a unified theoretical framework, using the basic constructs of evolution and revolution as a conduit for this theoretical integration.

8.2 Organizational Change Motors

8.2.1 Organizational Change Motor Typology

The previous two chapters introduced, among other concepts, the dichotomy of evolutionary and revolutionary change. It bears repeating that in this context, evolution refers to the gradual shifts in an organizational configuration over time, a process that is punctuated by instantaneous revolutionary changes.

While scholars tend to agree that organizations go through both evolutionary and revolutionary change over the course of their development, the literature on the reasons why organizations develop in any particular manner has been inconsistent. A number of scholars have proposed sets of stages through which organizations will inevitably progress throughout their lifetimes (e.g., Burgelman and Sayles 1986; Flavell 1982; Kimberly and Miles 1980; Nisbet 1970; Rogers 1983), but with considerable disagreement about which model or models are accurate and which ones fail to represent real-world organizations. Others have tended to focus on the roles

© Springer International Publishing AG 2017   113
S.A. Matei, B.C. Britt, *Structural Differentiation in Social Media*, Lecture Notes in Social Networks, DOI 10.1007/978-3-319-64425-7_8

that internal and external dynamics play in actively shaping the trajectory of organizational change (e.g., Berger and Luckmann 1966, 1980; Brunsson 1982; Chakravarthy and Lorange 1991; Hegel 1812; Weick 1979), but again, with a great deal of disagreement about which factors might be important and which of them lack impact in reality.

Regardless of these disagreements between scholars, however, the broader claim that organizational growth progresses through a series of phases aligns with past research (e.g., Britt 2011, 2013). Moreover, when we apply this perspective to collaborative online production, such as that which takes place on Wikipedia, we might expect to detect a set of phase shifts at critical moments that separate and demarcate distinct phases. These phase shifts could represent "critical mass" moments in which the project reaches a threshold for growth that can only be surpassed by transitioning into a new structural form, or they may instead indicate the community's response to some coincidental factors that pushed it into a new way of conducting its daily practices. In either case, past research suggests that organizational configurations do in fact undergo significant, revolutionary changes of this sort—we simply do not know the reasons for those changes.

With both the importance of organizational change and the historical conflicts between change scholars in mind, we turn our attention to Van de Ven and Poole (1995), who offered one of the most comprehensive typologies of organizational change to date. In constructing this framework, Van de Ven and Poole sought to synthesize the four primary classes of change theories—life-cycle, evolutionary, dialectical, and teleological—into a single typology describing the forces, or "motors," that drive change.

Let us first briefly summarize these four classifications. Theorists of life-cycle change argue that organizations develop like organisms, progressing through a predetermined linear sequence toward an inevitable conclusion, with little more flexibility than a railroad car moving down a set of train tracks. Evolutionary theories—not to be confused with the "evolutionary change" that is dichotomized with "revolutionary change"—instead focus on the environmental forces that promote and maintain the three key evolutionary processes of variation, selection, and retention among different organizations and organizational forms. Dialectical theories highlight conflicts as they emerge between actors with competing goals or values as well as the ways in which the resolution of those conflicts may establish new norms within the organization and within the interorganizational context. Finally, teleological theories consider the presumed "final state" of the organization and use developments that push the organization toward this end result as the standard for judging change, even though this destination itself may change over time such that the organization never actually reaches a point of stagnation, stability, or finality.

Among these, the evolutionary and teleological change theories rank as the most prominent in the literature (Phillips and Duran 1992), but all four are sufficiently established to stand apart from the others (see Levy and Merry 1986). Others have proposed additions to Van de Ven and Poole's typology, especially the social cognition and cultural approaches to change (see Kezar 2001), but there is still significant doubt about whether such additions would offer a significant improvement upon the existing typology or merely clutter an otherwise parsimonious explanatory scheme.

Table 8.1 Scope and determinism dimensions of organizational change motors		Determinism	
		Inescapable	Actor agency
Scope	Single unit	Life-cycle	Teleological
	Interactions	Evolutionary	Dialectic

Importantly, Van de Ven and Poole (1995) did not envision these four theories as mutually exclusive perspectives that necessarily compete against one another, but as classes of potential "motors" for a given organization. They argued that an organization may be driven by any number of these four types of motors and that the various motors may struggle against each other or play complementary roles in any given context.

With that in mind, these four classes of motors vary across two dimensions: scope and determinism (see Table 8.1). Evolutionary and dialectic motors arise from the interactions between multiple entities, while motors associated with the life-cycle and teleological theories focus on the way in which a single unit naturally changes over time solely based on its own natural progression or the direct influence of a single party on the organization. Likewise, the life-cycle and evolutionary frameworks treat certain changes as inescapable, whether they are internally defined trajectories imposed upon the organization by those who first established it or environmental forces that the organization is powerless to combat, while the dialectic and teleological frameworks prioritize the agency of individual actors to challenge one another and to spur the organization toward its ultimate purpose, respectively.

In other words, evolutionary motors are the exogenous forces that act upon the organization, or the environmental influences to which the organization must adapt in order to survive and to thrive. A life-cycle motor is, in essence, the natural progression of the organization along the largely predestined course that was laid when it was first formed. While life-cycle motors emerge from within the organization itself, its members cannot easily reshape or resist those motors, much like humans can do little to manipulate their own progression through their respective life cycles.

It is therefore largely impossible for organizations to escape either of these forces. An organization can, in theory, try to resist either type of motor, but it does so at its own peril. An organization that ignores the larger environment in which it is embedded is likely to quickly become irrelevant and be abandoned. Likewise, an organization that resists its own need to grow and evolve as its makeup changes is apt to stagnate and collapse. Either way, the organization must change in order to survive. With that said, the members of an organization may be able to exert some limited control over certain elements of the change, coaxing the organization in one direction or another, but the change itself is largely inevitable.

Dialectic and teleological motors, on the other hand, both represent forces that stem from the internal activities of their members rather than circumstances outside their control. Dialectic motors emerge from dialogue between individuals within the organization. When organizational members disagree upon the direction in which they feel the organization needs to progress, they may clash with one another, struggling as opposing forces that both aim to move the organization forward in their own ways. Often this disagreement manifests as a conflict between change and the status quo, and it may result in the organization shifting toward a synthesis of the two competing ideas, a victory for one ideology over the other, or an entirely different alternative direction.

In contrast, teleological motors come from the force that a single unit imposes directly and deliberately on the organization. For instance, when a manager or managerial unit enacts a new policy or standard set of procedures, it constitutes a manipulation of the organization itself, in contrast with a dialectic process that generally works between fellow organizational members within the confines of the existing organizational framework. Teleological changes are much more fundamental exertions of an individual's will upon the organization than a mutual, collaborative process of reshaping the community, and these more deliberate changes to an organization are often further observed and adjusted as time goes on.

It should be noted that regardless of whether internal or external factors are identified as important motors for change in any particular organization, including Wikipedia, the result would not necessarily imply mere technological determinism. Even if external factors are particularly important in the development of the organization, it is quite possible that certain types of individuals tend to play key roles in organizations embedded in different contexts, such as the unique combination of users who have a great affinity for the developing cyberculture and who take a special interest in online organizations (see, for instance, Escobar 1994; Lévy 1997; Turner 2006)—a matter of personal agency in determining the internal organizational dynamic. Likewise, we may conjecture that the environment in which an organization resides offers different external restrictions and boundaries than other analogous environments, and the organization must chart its course differently in order to navigate this environment—again, implying that the organization has its own will instead of being passively pushed about like a sailboat adrift at sea.

All told, there is no inherent preference for any one change motor type over the others, as different motors are applicable to different situations. Rather, Van de Ven and Poole's (1995) framework is a heuristic taxonomy that was devised to help future researchers discern what organizational changes occurred at a given point in time, how they were manifest, and why they occurred in the observed manner. Importantly, this approach further provides us with a framework to describe the ways in which organizations create and manage their own social orders as well as how their members socially differentiate themselves over time (see Chap. 2).

With that said, it is worth recognizing that although no single type of change motor is necessarily "best" or "worst" across cases, it is quite possible that certain motors might be more or less useful in different organizational contexts. As such, there may be benefits for organizations whose leaders tailor organizational development by facilitating the influence of some motors while limiting the impact of others at different points in time, particularly in terms of the rate, direction, and perceived forces behind a given organizational change. This issue is discussed in more detail near the end of this chapter.

8.2.2   Configurational Change Motors

Taking the literature on both organizational configurations and organizational change motors into account, once the key evolutionary periods and revolutionary moments of configurational change have been identified, those points in time may be explored to determine why the organization developed in the way we have observed. In other words, we can assess what motors drove particular configurational changes by using the formal evolutionary and revolutionary changes to connect the heuristically defined configurations with the substantive evolutionary, life-cycle, dialectical, and teleological motors that caused shifts between them. Thus, by aligning the observed configurational changes with the underlying change motors that drove the observed organizational transformations, we may intertwine the two domains and uncover the primary configurational change motors that acted to push the organization from one organizational configuration to the next.

As for the particular methods used to identify organizational change motors, an extensive range of approaches may be used; a comprehensive list of potential methodologies is beyond the scope of this chapter. By and large, however, these techniques tend to be primarily qualitative examinations of interpersonal interactions along with organizational and public documents, with a considerable amount of subjective interpretation, particularly when comparing motors against one another to determine which ones were the more likely causes of a particular configurational change. For instance, when examining a particular revolutionary change, care must be taken to isolate internal and external forces that might reasonably have the observed effect, which coincided with the moment of revolution itself and whose lack of influence at other points in time can be explained by other factors (such as the motor's absence or other mitigating motors).

Importantly, the configurational change itself may serve as a litmus test for a motor's validity with respect to the change. If, for example, the revolutionary moment described above results in the organization moving from a relatively free-form configuration like the simple structure toward a more rigid machine bureaucracy or divisionalized configuration, then it is reasonable to ask how or why an observed motor would result in the organization becoming more bureaucratic. If we are looking for the motors that explain a particular configurational change, it makes sense to use the configurational change itself to evaluate the validity of those motors as explanatory mechanisms, thereby connecting the frameworks of organizational configurations and organizational change motors on a methodological as well as a theoretical level.

As a whole, these conceptual interconnections serve to interweave the otherwise distinct theoretical frameworks of organizational configurations and organizational change motors by using the basic change dichotomy of evolution and revolution as a conduit between the two areas. This unification will allow scholars and practitioners alike to take advantage of the heuristic parsimony of Mintzberg's Configurational Array (1979, 1989) to quickly diagnose organizational configurations, while also recognizing that those configurations, far from static, are subject to change over time, which we may explore on a more nuanced level via Van de Ven and Poole's (1995) organizational change motors. In short, this synthesis blends a straightforward configurational diagnostic instrument with a deeper consideration of the causal factors driving configurational changes, thereby enriching our understanding of both theoretical domains.

8.3   Practical Implications

This chapter serves to complete a new theoretical framework that unites existing explanations of the ways in which organizations proceed through evolutionary and revolutionary change, the various configurations that organizations adopt, and the key organizational and environmental motors that fuel organizational growth. The digital age offers countless opportunities for individuals to contribute to an array of organizations and help to reshape those communities from the inside, making a deeper understanding of how organizations and their configurations change over time all the more important.

Both Mintzberg's Configurational Array (1979, 1989) and Van de Ven and Poole's (1995) organizational motors were designed to unify many competing fragmented theories into a single explanatory mechanism. Similarly, this chapter fuses two disparate research domains, bringing together ideas and scholars that might otherwise remain apart and aiding in the development of unified theories to explain online organizational change and social structural development.

All told, this new synthesized framework offers a fundamentally enhanced approach to examine the long-term transformation of organizations and to understand the motors that form and reconstitute organizational configurations throughout their development. Of course, one important question remains: How can we use these findings in the real world? More specifically, what if practitioners were to use this unified framework to guide decision-making? What can we learn from this new theoretical model?

8.3.1   Identifying Existing Informal Configurations

The first lesson to apply stems from the simple fact that communication patterns rarely align with formal organizational structures. This is just as true in any company or nonprofit organization (Kraut et al. 1990) as it is on Wikipedia, where there is indeed a hierarchy that features a small group of administrators, but for which editing and communication patterns hardly conform to that hierarchy. Certainly, some misguided corporate managers may demand that their employees communicate only with coworkers to whom they are officially connected on a formal organizational chart. Such efforts not only stymie the organization and its members; they also generally fail.

Broadly speaking, there will inevitably be conversations that occur outside these restrictions—and, in many cases, such "water cooler" conversations spawn the most productive, innovative, and collaborative work in the organization. Therefore, it is in the best interest of any manager, community organizer, or moderator to examine the communication that is happening beyond the formal organizational structure. These interactions constitute the emergent structure of the organization—the form toward which it is gravitating on its own, regardless of any official mandates otherwise—which may very well be a configuration that would operate more effectively than the formal one that the creators or managers of the organization imposed. If that is the case, and if the organization could benefit by being reorganized around this emergent structure, then it would certainly behoove an intelligent manager to be able to identify that configuration. This is where Mintzberg's typology comes into play.

Of course, as we have discussed at length, a snapshot does not offer a sufficient picture to define an organization. As such, the unofficial communication network, or the emergent structure, should be assessed over time in order to see how, indeed, it is "emerging." Doing so will allow an intelligent manager or community creator to assess the trajectory of this emergence—Is the unofficial network becoming more or less similar to the formal organizational chart? Toward what configuration archetype is it moving?—as well as the various motors that might be driving this configurational change.

Understanding these issues will allow a smart organizational "instigator" a means to assess the communicative health of his or her organization, including the relevance of the formal communication structure in the actual day-to-day activities of the organization members.

8.3.2 Influencing Informal Configurations

With that in mind, it stands to reason that if an organization's creator, moderator, or manager can gain extra insight into his or her organization by assessing the configurations represented by its informal communication networks, then it might be worthwhile for the leaders or moderators to influence those informal configurations in order to serve organizational interests in the long run.

It should first be noted that such deliberate efforts are not always appropriate, and in fact, the emergent communication structure is often superior to one that a moderator or leader might impose. After all, when users or organization members go out of their way to interact with others with whom they are not "officially" connected, it is often out of necessity (see, for instance, Krackhardt and Stern 1988), to take advantage of a unique opportunity (Kraut et al. 1990), to obtain a more diverse perspective (Burt 1995), or to otherwise improve their work and the company itself. Why, indeed, should such communication be discouraged?

Further, even if the informal configuration is developing in a manner that greatly diverges from the formal communication network, this is also not necessarily a problem. It could be that an internal or external need or opportunity is driving the shift, or the emerging configuration might merely be part of the next stage in the organization's life cycle—perhaps it has outgrown its old form, and the formal organizational chart simply has not caught up to the new communication network.

For that matter, there is likely some benefit in allowing communication beyond that of a rigidly defined formal organizational chart, whether in online communities or in businesses. In such contexts, members or employees who feel free to interact with their peers as they please, rather than being ordered to interact with some individuals while remaining forbidden to speak to others, will likely be happier, more creative, more dedicated, and generally more productive as a result. Beyond this, an overly controlling manager who compels especially rigid patterns of interaction can sometimes push subordinates away, giving them cause to deliberately interact with those outside their assigned cadre simply as a momentary escape. Consequently, when the informal communication network wildly diverges from the formal configuration, it may be a sign that management has been too restrictive and that it may be worthwhile to allow employees a bit more freedom and flexibility than they have previously been given.

Yet, there may be some cases, such as in business organizations that extend online, in which a manager does need to exert a degree of control over the developing informal communication configurations. As an extreme example, if a few mutinous employees are fueling unrest within the company, then they may need to be removed and, more broadly, the organization may need to be reconfigured in order to prevent future cliques of dissatisfied subordinates from wreaking similar havoc in the future. More commonly, communication that occurs outside the bounds of an organizational chart may distract employees or lead to the spread of rumors and other disinformation more quickly than it can be contained or counteracted.

Either way, and across various online and offline organizations, a smart leader, creator, or moderator will consider the alignment of formal and informal communication networks with one another and will evaluate the manner in which they are changing over time, whether they are growing in tandem or in competition with one another. A leader who is flexible in considering alternative organizational configurations will be able to adapt his or her organization to suit its ideal form—perhaps the one that its members are already trying to adopt on their own—rather than being tied to an outdated, historical structure. In other words, forward-thinking leaders can use this approach to seek opportunities for positive change rather than being married to the past.

Sadly, though, as Crampton et al. (1998) noted almost two decades ago, managers rarely take an active role in guiding or even swaying the informal communication networks of their organization's members, especially in more formal settings, which may straddle offline and online environments. Little has changed since their initial study. The importance of informal configurations has largely been neglected in orchestrating complex organizations, especially collaborative ones, with leaders neglecting to use their influence to affect or even monitor such activities. Left unchecked, a detrimental informal configuration can undermine the entire group's activities, driving member discontent and internal strife. When this happens, groups suffer and slowly die.

8.3.3 Inciting Formal Configurational Change

As suggested above, there are many benefits in terms of both user satisfaction and effective collaboration to be derived from allowing organizations of all kinds—and especially online collaborative ones—to develop along a path set forth by their members, especially if doing so leads to member-inspired innovation. This is just as true of formal organizational configurations as it is of any informal communication networks.

However, it is sometimes appropriate for leaders and organizers, formal or informal, to guide an organization themselves in order to deliberately shape its growth rather than leaving it up to the collective judgment of their less influential peers. This frequently occurs on Wikipedia, and such activities can be used as exemplars for the development and maintenance of collaborative online organizations. After all, emergent communication structures in such environments sometimes develop out of self-interest rather than with the organization's goals in mind, and even well-meaning beginning or intermediate users or members often lack the wide-ranging perspective that a community leader or creator with a bird's-eye view of the organization has in orchestrating interaction patterns. With that in mind, a leader or collaborative community organizer may wish to influence two interrelated elements of the formal organizational structure: the configuration toward which the organization is moving, and the motors that are driving it there.

The rationale for the first element is obvious. In order for an organization to continue to survive and thrive, it needs to be structured such that it can operate in an ideal manner. Naturally, then, the configuration that the organization adopts is important. A global social media platform, for instance, likely would not be well served by adopting a simple structure, nor would a divisionalized configuration be appropriate for a fledgling Internet forum that is just beginning to grow. Online organizational leaders should consider the kind of interactions in which their members are engaging and what interactions are actually necessary in order to foster success, be it in terms of entertainment, content production, social networking, or something else entirely. Notably, the configuration of the existing informal communication network might serve as an important factor for such decisions, though it is unlikely to be the sole determinant.

On the other hand, it is not as obvious why a leader would want to be concerned with the organizational motors that drive the change between configurations, nor, for that matter, what control the leader might have over them. After all, if the change is a positive one, why should anyone care what causes it?

This is a dangerous sentiment. To take a lesson from the business world, history is littered with cases in which well-meaning managers drove their companies to ruin by trying to enact a positive change but doing so ineffectively. We may easily expect the same to be true of online organizations. As a simple illustration of this phenomenon, in any given business, few things scare and anger employees more than having change forced upon them. In general, most people tend to fear change anyway, and this is all the more true when one does not have a voice in the process.

When the environment in which we live changes around us, without any input or consent on our part, it makes it difficult for us to even function, let alone thrive as we once did. Maybe it is, indeed, a good idea to combine departments, to change managerial lines, to reassign roles, and so forth. Nonetheless, if the change is thrust upon employees, then they are likely to resist, opposing the new policies and defaulting to their existing informal communication networks, their previous formalized hierarchy, and their long-established ways of interacting. Such dissent would certainly stymie the organization as a whole, and it could ultimately turn any positive growth that might have been envisioned into the death knell of the business.

Yet many managers nonetheless rely on sweeping mandates to direct their employees to behave in a desired manner. There are isolated instances in which it is indeed necessary to enact major policy changes without consulting with one's subordinates—particularly to stifle crises before they can fully erupt—but if your organization is generally healthy and your aim is simply to make it even stronger, there is little reason to do so.

In short, employees adapt best to change when they respect and trust its initiator or feel that they have a degree of control over the forces driving the organizational shift. More broadly, the lesson is that the motors driving a given change matter just as much as whether or not the change itself is well founded.

Therefore, in both online and offline organizations, it is worthwhile for a smart leader to consider the motors that should be fostered and those worth tempering in the process of inciting an organizational change. For instance, if you hope to make your formal organizational configuration more closely resemble the existing informal communication network, then it is easy to allow internal motors to drive the change: just gradually relax the formal restrictions and let the organization transform naturally. Perhaps, on the other hand, it is worth pursuing a more dramatic change in order to help the organization progress beyond a point of critical mass, in which case you will want to highlight the life-cycle motors at play that necessitate the change. In other contexts, it might be worth breaking existing patterns of internal communication in order to allow external forces to play more of a role, thereby making it easier for the organization to adapt to its environment.

The point here is that the driving forces—or, at least, the perceived motors—behind a change are sometimes even more important than whether or not the change itself is a good idea. Yes, a good leader will want to help guide his or her organization in a positive direction, and such an individual may have the vision to see what configurations would be most useful for the future. Yet the motors that drive this change must also be given due attention, as members' sentiments about the causes behind a change are a prerequisite for reacting positively to the change itself. In the end, rather than moving the community forward, blindly attempting to compel a constructive configurational change is likely to instead backfire and send the organization into decline. In this respect, the process of achieving a change is just as important as its product.

8.4 Conclusion

In closing, the focus of the last three chapters was on providing much more specific explanatory mechanisms for the emergence of social structures. Notably, not all organizations are the same, and not all phases through which an online collaborative organization progresses are triggered by similar motivations or socio-technological mechanisms. Therefore, our theoretical approach acknowledges the variety of configurations that an organization may adopt and the range of motors that may drive changes in those social structures. This allows researchers to properly assess the trajectories through which organizational growth and development may occur in different contexts, and it provides practitioners with the means to guide their own groups along a healthy trajectory, while simultaneously avoiding the pitfalls of a misguided one-size-fits-all conceptualization of online and offline social structures.

The primary goal of these three chapters was to provide a clear and detailed explanatory framework along with a complex, concrete, and operationalizable method for discerning organizational structuration and change. The payoff is not merely theoretical, as there is also a practical dimension to our propositions. Above and beyond measurement and diagnosis, the method offers valuable insights about how formal and informal collaborative processes can be harnessed to grow and maintain online collaborative spaces. All told, this new synthesized theoretical framework offers a fundamentally enhanced approach to examine the long-term transformations of organizations and to understand the motors that shape, affect, and transform organizational configurations throughout their development, yielding a theoretical lens that offers valuable insights to scholars and practitioners alike.

References

Berger PL, Luckmann T (1966) The social construction of reality: a treatise in the sociology of knowledge. Anchor Books, New York

Berger PL, Luckmann T (1980) The social construction of reality: a treatise in the sociology of knowledge, 1st Irvington edn. Irvington Publishers, New York

Britt BC (2011) System-level motivating factors for collaboration on Wikipedia: a longitudinal network analysis. Thesis, Purdue University

Britt BC (2013) Evolution and revolution of organizational configurations on Wikipedia: a longitudinal network analysis. Dissertation, Purdue University

Brunsson N (1982) The irrationality of action and action rationality: decisions, ideologies and organizational actions. J Manage Stud 19(1):29–44

Burgelman RA, Sayles LR (1986) Inside corporate innovation: strategy, structure, and management skills. Free Press, New York

Burt RS (1995) Structural holes: the social structure of competition. Harvard University Press, Cambridge

Chakravarthy BS, Lorange P (1991) Managing the strategy process. Prentice Hall, Englewood Cliffs

Crampton SM, Hodge JW, Mishra JM (1998) The informal communication network: factors influencing grapevine activity. Public Pers Manag 27:569–584

Escobar A (1994) Welcome to Cyberia: notes on the anthropology of cyberculture. Curr Anthropol 35(3):211–231

Flavell JH (1982) Structures, stages, and sequences in cognitive development. In: Collins WA (ed) The concept of development: the Minnesota symposia on child psychology. Lawrence Erlbaum Associates, Hillsdale, pp 1–28

Hegel GWF (1812) Wissenschaft der logik. Johann Leonhard Schrag, Nürnberg. English edition: Hegel GWF (1998) Hegel's science of logic (trans: Miller AV). Prometheus Books, Amherst

Kezar AJ (2001) Understanding and facilitating organizational change in the 21st century: recent research and conceptualizations. Jossey-Bass, San Francisco

Kimberly J, Miles R (1980) The organizational life cycle. Jossey-Bass, San Francisco

Krackhardt D, Stern RN (1988) Informal networks and organizational crises: an experimental simulation. Soc Psychol Quart 51(2):123–140

Kraut RE, Fish RS, Root RW, Chalfonte BL (1990) Informal communication in organizations: form, function, and technology. In: Oskamp S, Spacapan S (eds) Human reactions to technology: the Claremont symposium on applied social psychology. Sage, Beverly Hills, pp 145–199

Lévy P (1997) Cyberculture. Odile Jacob, Paris. English edition: Lévy P (2001) Cyberculture (trans: Bononno R). University of Minnesota Press, Minneapolis

Levy A, Merry U (1986) Organizational transformation: approaches, strategies, theories. Praeger Publishers, New York

Mintzberg H (1979) The structuring of organizations: a synthesis of the research. Prentice Hall, Englewood Cliffs

Mintzberg H (1989) Mintzberg on management: inside our strategic world of organizations. Free Press, New York

Nisbet RA (1970) Developmentalism: a critical analysis. In: McKinney J, Tiryakin E (eds) Theoretical sociology: perspectives and developments. Meredith, New York, pp 167–206

Phillips R, Duran C (1992) Effecting strategic change: biological analogues and emerging organizational structures. In: Phillips RL, Hunt JG (eds) Strategic leadership: a multiorganizational-level perspective. Quorum Books, Westport, pp 195–216

Rogers E (1983) Diffusion of innovations, 3rd edn. Free Press, New York

Turner F (2006) From counterculture to cyberculture: Stewart brand, the whole earth network, and the rise of digital utopianism. University of Chicago Press, Chicago

Van de Ven AH, Poole MS (1995) Explaining development and change in organizations. Acad Manag Rev 20:510–540

Weick KE (1979) The social psychology of organizing, 2nd edn. McGraw-Hill, New York

Chapter 9
Wikipedia Evolution: Trends and Phases

9.1 Introduction

In the previous chapters, we presented a theoretical framework to explain the emergence and transformation of complex organizational configurations from an integrated perspective. We suggested that structural differentiation is an early and constitutive process by which online social organizations emerge, grow, and become stable. In this chapter we explore some empirically measurable trends on Wikipedia that more concretely reveal structural differentiation through the lens of organizational configurations.

As a reminder, structural differentiation can take several forms or organizational configurations, each of which may be identified by the combination of our five different effort distribution and network dimensions (see Chap. 7). In other words, organizational configurations are observed in terms of collaborative ties and associated social graphs of interaction along with the division of labor among collaborators, and those configurations are, in turn, used to explain the growth and maturation of collaborative projects such as Wikipedia. This provides a unique frame of reference to study online production communities of this sort, whose forms naturally emerge from user activity rather than being imposed by a top-down management scheme and deliberate organization design.

In this chapter, we analyze data across the five dimensions of inbound and outbound degree centralization, betweenness centralization, assortativity, and entropy to track the ways in which Wikipedia has changed over time. We also propose a statistical method that may be used to determine the moments at which internal and external phenomena triggered revolutionary shifts in an organization's development, and we subsequently use that method to identify transformative changes in the social structure of the online group that built Wikipedia.

© Springer International Publishing AG 2017

S.A. Matei, B.C. Britt, *Structural Differentiation in Social Media*, Lecture Notes in Social Networks, DOI 10.1007/978-3-319-64425-7_9

9.2   Configurational Phase Shifts: Criteria and Significance

Recall, first and foremost, that structuration is expected to progress through certain phases over time. In Part I of this volume, structuration was examined through only one dimension: entropy. In this chapter, we consider changes in entropy as well as four other collaborative network factors: three types of centralization (betweenness, inbound degree, and outbound degree) and assortativity.

As was explained in Chap. 7, inbound degree centralization describes the differentiation in the extent to which some individuals attract collaborative partners, while outbound degree centralization refers to differentiation in the extent to which some members pursue collaborative ties with others. In other words, this is the difference between seeking others (outbound) and being sought after (inbound). Note that in this study, the direction of collaborative ties is determined by the order in which individuals edited a given article, as described in Appendix A.

Betweenness centralization refers to the extent to which certain individuals in a collaborative network act as bridges, creating "shortest paths" of collaboration that unite otherwise disparate partners and groups. Finally, assortativity is the tendency of individuals to associate with those who are similar to them on a given dimension. In this context, this similarity emerges in the form of similar productivity levels among coeditors. High assortativity would indicate that individuals who are more productive are concentrated in a local network of highly productive contributors, while disassortativity would instead suggest that top contributors prefer to engage with fledgling contributors rather than fellow elites.

These four measures reflect four core organizational processes. Inbound centralization represents collaborative attractiveness, outbound centralization signifies collaborative extroversion, betweenness centralization provides a summary of communication flow and coordination, and assortativity is our operationalization for the concept of partner choice tendencies (for further details, see Chap. 7). These complement entropy, our measure of structural order, to jointly produce a comprehensive view of an organization's principal attributes and processes.

We use the five dimensions together to detect the "breakpoints" that occurred in the development of Wikipedia's collaborative evolution and to paint a more nuanced image of its changes over time. Breakpoints appear where "revolutionary" changes can be detected via statistical analysis, including those signifying transitions between configurations. By exploring the tendency for organizations like Wikipedia to move from one configuration to another, we also gain precious insight into the natural inertia of organizations in general and of online social collaborative spaces in particular.

The synthesized theory established in Chaps. 7 and 8 is complemented by the methodological perspective and analysis presented in this chapter. In Sect. 9.3 we provide a primer on the statistical requirements for a suitable algorithm to detect development thresholds, then in Sect. 9.4 we will introduce a stepwise segmented regression analysis that satisfies those needs. If you are interested in a more technical overview of the algorithm, please refer to Britt (2013, 2015) and Appendix A for further information.

Afterward, we will use this method to identify the theorized periods of evolution and moments of revolution, and with them, the activation of various phases of organizational change over time. This will showcase the points of transition from one organizational configuration to another, which is essential for our final description and interpretation of the phases in the development of structurally differentiated online collaborative groups.

In effect, in this chapter we upgrade the analysis presented in Part I of this volume from a one- to five-dimensional investigation of structural differentiation. Our goal is to identify the specific organizational configurations that emerge and succeed each other in online production spaces, such as that surrounding Wikipedia. Of particular importance is detecting the emergence of a final form of structural differentiation, which the analyses earlier in this volume suggested might emerge as an adhocratic order.

9.3 Existing Breakpoint Detection Algorithms

The analytic portion of this chapter will focus on determining breakpoints in five collaborative dimensions, which jointly map onto the theoretical model developed in the previous three chapters. The dimensions reflect long-term measurements of the set of factors (inbound degree centralization, outbound degree centralization, betweenness centralization, assortativity, and entropy) introduced in Chap. 7 to describe organizational configurations. In other words, a set of time series was created, with one such series corresponding to each of the five key factors.

The time series, which are described in more detail in Appendix A, were constructed on the basis of a quasi-complete network of collaborative ties between Wikipedia users. The ties in this network were derived from coeditorial interactions. A user was considered to be connected to another user if both individuals edited the same article, as doing so suggested that they may have been engaging with one another or, at the very least, with one another's thoughts and ideas. The more significant the revisions and the closer they were together in time, the stronger the weight of the edges, indicating an increasing likelihood that the users were indeed engaged with each other's work. Such edges additively grew in degree over multiple sets of revisions and multiple articles, strengthening over time, with the result being a probabilistic network representing the relative likelihood of any given pair of users holding a significant collaborative relationship.

The network was examined based on the manner in which it grew on a weekly basis. A single network graph was constructed corresponding to each week of Wikipedia's development over its first 9.5 years, with any given graph containing all of the edges that resulted from all revisions made during or prior to the week in question. This resulted in the generation of almost 500 network graphs, each of which added the edges that resulted from one additional week of revisions to the graph that came before it.

Afterward, these networks were all analyzed in order to measure the values of our five key metrics: inbound degree centralization, outbound degree centralization, betweenness centralization, assortativity, and entropy. All told, this approach yielded five distinct time series that each offered a description of one organizational factor related to Wikipedia's development over its first 9.5 years.

Further, we have already described the importance of distinguishing between prolonged evolutionary shifts and nigh-instantaneous revolutions, so these must be identified within the time series. For this purpose, we refer to the idea of "breakpoints" that, when apparent in time series data, present themselves as changes in the path of the plotted line. In more concrete terms, a breakpoint may consist of a change in the line's intercept, slope, or an exponential term. In this sense, a breakpoint is a tangible representation of a revolutionary change, as evidenced by a dramatic, instantaneous shift in one of the major factors being examined.

The core idea behind this approach is that from a measurement standpoint, only sharp, abrupt changes matter, since they capture breakpoints in the evolutionary process. In truth, while we are also interested in periods of evolution, only the revolutionary moments must be pinpointed—after all, if we know when a finite number of revolutions occurred to fundamentally change the system, then the periods between those dramatic shifts must instead constitute evolution or stability. So if we can identify the revolutions, then we have by definition denoted the evolutionary periods as well. These revolutions, when apparent in time series data, present themselves as breakpoints in which path of the plotted line changes. In terms of the graph itself, a breakpoint may consist of a change in the line's intercept, slope, or an exponential term.

Breakpoints are common reference points in quasi-experiments, such as those conducted in pharmacology (Wagner et al. 2002). In these studies, some metric related to a subject's health or aptitude is measured over a period of time. At a pre-specified time, a treatment (such as a new medication) is applied to the subject, and measurement continues thereafter as the treatment continues. The moment at which the treatment was applied is deemed the breakpoint, and separate regression equations can be fitted to the line segments that occurred before and after the breakpoint. If there is a statistically significant difference between the parameters of the two line segments, then the treatment is deemed to have had an effect accordingly.

It is easy to see how breakpoints, then, signify "revolution," in this case indicating the application of a treatment. Yet, in quasi-experiments, the location of the breakpoint is known in advance: it's the moment at which the treatment was initiated. This is not the case in studies of emergent processes, such as collaborative behavior on Wikipedia, for which we would like to determine when moments of revolution *naturally* occurred, without a researcher's deliberate intervention. So, while the importance of breakpoints remains, we need some means of identifying where those breakpoints are present and even how many breakpoints there actually are.

A number of existing methods promise to identify breakpoints in order to serve this need. Unfortunately, each such method carries several significant limitations that mitigate its usefulness for studies of this sort (see Britt 2015 for a more detailed overview). For instance, the steepest descent and gradient descent models, as well as the Newton, Gauss-Newton, and Marquardt algorithms commonly employed by the statistical package SAS, only work with continuous breakpoints, which may not be a realistic assumption for volatile online social systems in which the community may transition into a completely different form almost overnight. These approaches also presuppose a specific number of breakpoints, which cannot be known in advance when one is trying to detect revolutions between an unknown number of developmental stages.

Likewise, although Achim Zeileis' strucchange algorithm (Zeileis et al. 2003, 2010) might appear to offer more promise for fitting an unknown number of continuous and discontinuous breakpoints to real-world data, Britt (2013) found that this algorithm dramatically overfit data with substantial amounts of random noise, attributing an unrealistic number of breakpoints that merely fit error variance and generating comparatively mediocre fits to genuine revolutions. We would expect measurements of online communities to have exceptionally large degrees of error variance, and for growing or declining groups, such error variance is frequently of a heteroscedastic nature, which further confounds Zeileis' approach. As a whole, these phenomena render the strucchange algorithm wholly unsuitable for breakpoint detection in online communities.

All told, no previously existing algorithm provides an acceptable method to identify revolutions in an online community like Wikipedia, yet the ability to pinpoint these breakpoints is essential for assessing organizational growth over time. This need drives us toward the creation of a new statistical method that is especially suited for breakpoint detection in volatile time series data sets, such as those of online social systems.

9.4 Stepwise Segmented Regression Analysis

This new approach, called stepwise segmented regression analysis, draws from the principles of stepwise model selection as well as the traditional use of segmented regression analysis in quasi-experimental designs, offering a new way to detect breakpoints in volatile environments like those of online communities and other sites of human interaction. Its basis in regression analysis offers a degree of robustness against high-variance data sets and heteroscedasticity, and it is flexible enough to handle continuous and discontinuous breakpoints alike in raw data sets with no need for data transformation or smoothing.

Stepwise segmented regression analysis is, at its core, a modified form of segmented regression. Just as in a quasi-experiment, this analysis is designed to detect the differences, if any, between two or more adjacent line segments. These line segments may be modeled with intercepts, slopes, and any number of exponential terms, just as in typical regression analyses. The primary modification that the stepwise segmented regression algorithm applies is that every interval between any two adjacent data points is considered to be a possible breakpoint and stepwise model selection is used to select which breakpoints, if any, are statistically significant. This provides us with a straightforward way to detect social revolutions as evidenced by breakpoints of any type.

The basic procedure behind stepwise segmented regression analysis is as follows. First, a model is fit with only the basic terms defined by the researcher (such as an intercept, slope, and any higher-order regression terms). This set of terms can be considered the first level of a hierarchical regression, as they are never removed from the model regardless of significance level.

Second, each possible breakpoint is considered for inclusion in the model based upon a standard model selection threshold such as $\alpha = 0.15$. All terms associated with an interval between two points are considered. For instance, if the initial model included an intercept, a slope, and a quadratic term, then each potential breakpoint would also constitute intercept, slope, and quadratic terms representing the change in each parameter occurring at that breakpoint. A breakpoint consisting of all three terms would then be added to the regression model if and only if at least one of its three terms was statistically significant at the $\alpha = 0.15$ level. It should be noted that the reason for adding all three terms associated with a given breakpoint to the model, rather than just adding the statistically significant term or terms, is to avoid erroneously fitting breakpoints resulting from different parameters at consecutive or proximal data points when only one revolution has in fact occurred.

Third, if any breakpoints already in the model have all of their associated terms cease to meet the $\alpha = 0.15$ level of significance, then all terms associated with those breakpoints are removed from the model. To be clear, all terms related to a given breakpoint (intercept, slope, quadratic, etc.) are all added to the model together and deleted together; in the interest of avoiding spurious "extra" breakpoints in the model, individual terms are never added or removed in isolation.

Steps 2 and 3 then continue to repeat until no further breakpoints can be added or removed. The resulting model offers the starting position and initial trajectory of the parameter or parameters measured in the time series, the changes that resulted from any breakpoints (revolutions) identified, and the evolutionary trends observed in the periods between those breakpoints.

Importantly, some of our network measures, especially the centralization metrics, are not necessarily directly comparable across networks of different sizes (Freeman 1979). However, this is not a problem for breakpoint detection using stepwise segmented regression analysis. Let us assume, for instance, that a metric such as betweenness centralization is naturally predisposed toward a downward curvilinear trend as the number of members within a given organization increases. In that case, the "first level" of the regression would be used to model that curvilinear trend, while the stepwise procedure would capture any significant deviations from that trend—which are, indeed, the breakpoints that we would like to assess.

9.5 Breakpoints as Revolutionary Change

The theoretical model behind our analysis is premised on the idea that dramatic revolutionary changes in key time series variables (entropy, centralization of each kind, assortativity) indicate breakpoints or thresholds that mark transitions in core organizational configuration dimensions. When specific levels in these dimensions align in specific ways, the combination of factors signifies the presence of a corresponding organizational configuration. Likewise, when the values of these dimensions change, organizational configurations change as well.

To recount, the five dimensions are the structures of collaborative attractiveness and extroversion, communication flow, partner choice trends, and structural order. For each, there is a specific measure—structures of collaborative attractiveness are measured using inbound degree centralization, those for extroversion are assessed with outbound degree centralization, communication flow is reflected in betweenness centralization, partner choice trends are detected via assortativity, and structural order is measured with entropy. A full justification for these associations is presented in Chaps. 7 and 8.

We used stepwise segmented regression analysis to examine changes in these five factors, as measured for the Wikipedia collaborative project from 2001 to 2010. (For a complete description of the data set, see Appendix A.) To that end, our five system-level parameters, which jointly tapped into Mintzberg's (1979) five organizational configurations, were assessed over the 9.5-year period. Figure 9.1 indicates the relative level we would expect for each dimension within each configuration archetype. For example, when inbound and outbound centralization are high, betweenness centralization is low, assortativity is either negative or near zero, and entropy is somewhat high, the observed organization resembles that of a simple structure configuration.

The diagram and the values in the cells of Fig. 9.1 reflect the theoretical model for organizational change presented in Chaps. 7 and 8. We derived the expected values and their combinations from a series of axioms, presented below. The axioms summarize the theoretical argument behind each configuration, as presented in Chaps. 7 and 8, reflecting the essential characteristics of each configuration. Corollaries, inferred from the axioms, provide guidance for determining the expected levels for each dimension in each organizational configuration. The breakpoints,

	Inbound Degree Centralization	Outbound Degree Centralization	Betweenness Centralization	Assortativity	Entropy
	Structure of Collaborative Attractiveness	Structure of Collaborative Extroversion	Structure of Communication Flows	Partner Choice Trends	Structural Order
Simple Structure Entrepreneurial	Very High	Very High	Low	Zero/Negative	Somewhat High
Machine Bureaucracy	Somewhat High	Somewhat High	High	Positive	Low
Professional Bureaucracy	Low	Low	Low	Positive	Low
Diversified Divisionalized	Somewhat Low	Somewhat Low	Very High	Positive	Low
Adhocracy	Low	Low	Low	Zero	Moderate

Fig. 9.1 Measurement levels for Mintzberg's configurational archetypes

then, provide guidance for interpreting the movement of these measurable dimensions over time, including whether they meet or fail to meet the combinations of thresholds suggested in Fig. 9.1 at various points in time.

The process of identifying organizational change in this manner differs from that discussed in Chaps. 3 and 5 in that it is multidimensional and it aims to capture not mere evolutionary phases but complex configurational changes. Of course, the sequencing of the previously identified phases and of changes in the organizational configurations should and do overlap. Yet, these sequences take on a new, deeper meaning through the explication of these configurational changes.

Based on the literature and the organizational configuration concepts presented in Chaps. 7 and 8, we propose the following axioms and measurement corollaries for determining the contours of the organizational configurations and the breakpoints separating them:

9.5.1 Simple Structure or Entrepreneurial Organization

Axioms
1. Resembles the classic "mom-and-pop" store.
2. One leader directly manages a collection of loosely organized subordinates.
3. Work is distributed: due to small size, everybody needs to do his or her share.

Corollary
Given the small size of the group, the leader will attract many inbound and outbound collaborative ties. However, the work is distributed, which means that there will also be lateral connectivity. Entropy will be somewhat inflated, reflecting the limited control that the leader exerts over the work process and the diffuse nature of the organization. Assortativity will also be low, as the leader, by necessity, has to work with everybody else. Betweenness centralization will also be low due to the distribution of collaborative ties and the resulting free flow of information.

9.5.2 Machine Bureaucracy

Axioms
1. Power concentrated at the top of the organization.
2. Leaders impose a fixed hierarchy upon subordinates.
3. Compartmentalization: minimal intermixing among unrelated work groups.
4. Unified structure.

Corollary
A small group of leaders, which is itself layered into tiers of power and communication, attracts a somewhat high level of inbound and outbound connections compared with the other organizational members. Their dominance is lower on the

attractiveness/extroversion dimension than the simple structure archetype, however, due to the plurality of commanding roles and the number of "middle management" layers separating the top of the hierarchy from the bottom. The stratification in this structure, though, does generate a higher level of betweenness centralization, with the flow of information across the hierarchy funneled through a limited number of intermediate leaders. Assortativity is prevalent in this archetype, as the organizational stratification and the small size of the top group both encourage preferential connections between fellow leaders. At the same time, entropy in a machine bureaucracy is low, as such systems are structured and relatively top-heavy by definition.

9.5.3   Professional Bureaucracy

Axioms
1. Elites are increasingly layered and more likely to designate power to middle management and team leaders.
2. Top groups are increasingly distant and separated from the rest of the organization.
3. Organizational power is, to a certain degree, localized within working groups.
4. Organization is integrated along a nexus of power.

Corollary
In a professional bureaucracy, a great deal of control is ceded to the operating core at the bottom of the hierarchy, so organizational power is not nearly as concentrated among a small clique of leaders and behavior is not as formalized as it is within the more rigid machine hierarchy. This leads to low centralization across the board, as communication and collaborative activity are both more distributed than they are in a machine bureaucracy. Assortativity remains high, indicating that, as with the machine bureaucracy, leaders are more likely to interact with each other than with members of the organization found on the lower levels. Entropy also remains low, as the amount of structuration is significant.

9.5.4   Diversified Organization

Axioms
1. Diversified organizations are federated organizations.
2. Work is the product of collaboration between entities that share some of their prerogatives at the highest level yet maintain a large degree of autonomy within themselves.
3. Some central individuals serve as essential bridges between entities/departments.
4. Basic units (divisions) have their own level of structuration.

5. Structuration within and across entities is relatively high.
6. Organizational divisions separate different types of members, fostering preferential connections between those who are similar.

Corollary

Within a diversified organization, the structure is fragmented into a number of largely distinct divisions, each of which maintains its own structure and standards of activity. The organization's activities are somewhat decentralized, yet the leaders serve as an absolutely vital link between the disparate units. Consequently, inbound and outbound centralization are somewhat low due to the highly fragmented nature of the organization, yet the federated nature of the organization demands that the organization's leaders serve as the point of connection between the divisions, so the overall level of betweenness centralization is correspondingly very high. The high level of structuration necessarily leads to a low level of entropy; on the other hand, the divisions between functionally distinct groups result in an assortative mixing pattern among members.

9.5.5 Adhocracy

Axioms

1. Structural organization is moderate, driven by "functional" leaders.
2. Functional leaders are born, not made. They are defined by what and how much they do, not by who they claim to be.
3. Structure of leadership is simple: leaders do, followers follow.
4. Structure is organic; members depend on each other to a larger degree than in bureaucracy.
5. Structure is temporally stable, yet there is temporal churn of elite membership.

Corollary

Adhocracies feature an organic structure similar to that of the simple structure but with the flexibility to form self-directed project teams as needed, based on the specialized skills of their members, and without an entrenched leader heavily dominating the structure from the top. Yet, adhocracy is not to be confounded with a total lack of organizational structure. Its distinguishing feature is the flexibility of roles and attributes, which is supported by the members' ability and willingness to take over specific roles and abandon them as needed. As a consequence, the only distinguishing feature of adhocracy is a moderate level of entropy and structuration. Inbound and outbound degree centralization and betweenness centralization are low. More generally speaking, an adhocratic organization follows either a disassortative mixing trend—with individuals gravitating toward others who are different from themselves—or no form of preferential attachment at all. This aligns with our findings from Chap. 5, in which we found that associations between elites and near-elite members do not improve the chances of those individuals on the cusp of a functional leadership position successfully ascending to the elite group.

9.5.6   Classification of Archetypes

Figure 9.1 consolidates the corollaries by indicating, for each organizational configuration, the expected measurement levels of each of the five dimensions based upon these axioms and corollaries. It should be noted that this is a relative classification system in which parameter values are described only as they are expected to change between the various configurations, but not with precisely defined numbers or ranges. The parameter associations were laid out in this manner because mere classification is not the goal of this analysis—rather, the objective is to assess movement between the archetypes. This goal does not demand that each archetype be rigidly and quantitatively delineated, and the lack of a strict numeric definition for each archetype permits some flexibility in its application across contexts.

As shown by Figs. 9.2 and 9.3, the observed levels of the five variables at various points in time suggest some interesting interpretations. All variables except assortativity start high, suggesting a configuration resembling the simple structure archetype. Three of them (entropy, inbound degree centralization, and outbound degree centralization) feature several apparent inflection points throughout the 9.5-year period, while assortativity and betweenness centralization trend toward nonsignificance over time. This indicates a shift toward adhocracy. What are, however, the breakpoints around which we can more definitively say that the organizational configuration dimensions are shifting in a significant, dramatic yet concrete way? What should be the revolutionary change boundaries in this assessment? More precisely, at what critical breakpoints do organizational configuration changes occur?

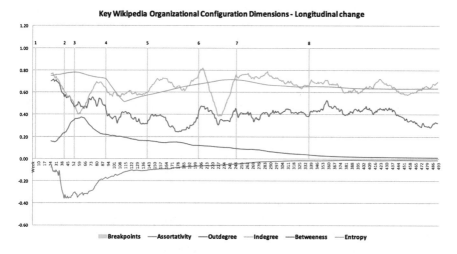

Fig. 9.2 Organizational change dimensions over time. This figure uses moving average values to smoothen periodic changes for the sake of interpretability. Moving averages calculated with a 24-week lag do not include the breakpoint occurring at week 7. Complete data can be found in Appendix A, Table A.2

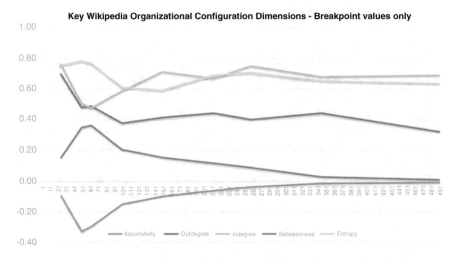

Fig. 9.3 A simplified version of Fig. 9.2 that includes only the values at each breakpoint, which are then connected by straight trend lines. This figure shows how values moved linearly from breakpoint to breakpoint. Vertices in the lines indicate breakpoints

9.6   Analysis

To determine these breakpoints, we ran five stepwise segmented regression analyses, one for each of the five parameters, for all weekly parameter values calculated across the 9.5-year period.[1] The first 3 and last 3 weeks of observations were removed from the data set due to issues with data completeness and parameter estimation, leaving 488 data points corresponding to weeks 4–491 of Wikipedia's existence for use in the analysis.

We identified 11 statistically significant breakpoints at the typical $\alpha = 0.05$ level after a Holm-Bonferroni correction (see Tables 9.1, 9.2, 9.3, 9.4, and 9.5). These 11 breakpoints corresponded to changes that were evident at the onset of weeks 7, 10, 42, 54, 92, 93, 142, 204, 207, 250, and 335 of Wikipedia's existence. Note that Chap. 5 only included those breakpoints resulting from our analysis of entropy, while this chapter also includes those from the time series for inbound degree centralization, outbound degree centralization, betweenness centralization, and assortativity.

As shown in Fig. 9.4, while many breakpoints suggest accelerating or decelerating transitions between configurations, only two breakpoints stand out as fundamental changes that dramatically altered the trajectory of the Wikipedia coeditorial configuration. Specifically, these are the shifts that occurred at weeks 54 and 92.

---

[1] The complete justification for this approach, including the rationale for conducting separate regression analyses instead of a single multivariate analysis, is provided in Appendix C.

Table 9.1 Regression model for inbound degree centralization

	Estimate	Std. error	t-value	p-value	
Intercept	6.112E^{-1}	2.686E^{-2}	22.753	$< 2\text{E}^{-16}$	*
Week	3.172E^{-4}	2.489E^{-4}	1.274	0.203	
Week^2	-4.798E^{-7}	4.844E^{-7}	-0.990	0.322	

*Denotes statistically significant variables ($\alpha = 0.05$)

Table 9.2 Regression model for outbound degree centralization

	Estimate	Std. error	t-value	p-value	
Intercept	7.674E^{-1}	5.671E^{-2}	13.533	$< 2\text{E}^{-16}$	*
Intercept* I (Week $\geq$ 142)	1.254E^{1}	2.381E^{0}	5.268	2.08E^{-7}	*
Intercept* I (Week $\geq$ 207)	-1.380E^{1}	2.391E^{0}	-5.771	1.41E^{-8}	*
Week	-5.608E^{-3}	1.799E^{-3}	-3.117	0.00194	*
Week* I (Week $\geq$ 142)	-1.473E^{-1}	2.764E^{-2}	-5.329	1.52E^{-7}	*
Week* I (Week $\geq$ 207)	1.583E^{-1}	2.762E^{-2}	5.734	1.74E^{-8}	*
Week^2	1.720E^{-5}	1.205E^{-5}	1.428	0.15397	
Week^2* I (Week $\geq$ 142)	4.290E^{-4}	8.008E^{-5}	5.357	1.31E^{-7}	*
Week^2* I (Week $\geq$ 207)	-4.541E^{-4}	7.919E^{-5}	-5.734	1.74E^{-8}	*

*Denotes statistically significant variables ($\alpha = 0.05$)

Table 9.3 Regression model for betweenness centralization

	Estimate	Std. error	t-value	p-value	
Intercept	2.029E^{-1}	1.799E^{-2}	11.274	$< 2\text{E}^{-16}$	*
Intercept*I (Week $\geq$ 54)	1.297E^{-1}	1.939E^{-2}	6.689	6.20E^{-11}	*
Week	-5.602E^{-3}	1.435E^{-3}	-3.904	1.08E^{-4}	*
Week*I (Week $\geq$ 54)	4.201E^{-3}	1.436E^{-3}	2.925	3.61E^{-3}	
Week^2	2.062E^{-4}	2.457E^{-5}	8.390	5.37E^{-16}	*
Week^2*I (Week $\geq$ 54)	-2.047E^{-4}	2.457E^{-5}	-8.328	8.49E^{-16}	*

*Denotes statistically significant variables ($\alpha = 0.05$)

Table 9.4 Regression model for assortativity

	Estimate	Std. error	t-value	p-value	
Intercept	-2.802E^{-1}	1.697E^{-2}	-16.513	$< 2\text{E}^{-16}$	*
Week	1.426E^{-3}	1.569E^{-4}	9.089	$< 2\text{E}^{-16}$	*
Week^2	-1.846E^{-6}	3.048E^{-7}	-6.057	2.78E^{-9}	*

*Denotes statistically significant variables ($\alpha = 0.05$)

These appear to delineate intervals that align, based on the values of the five dimensions being measured, with three distinct organizational configurations: simple structure, machine bureaucracy, and adhocracy.

More specifically, in the beginning of Wikipedia's existence, the social system featured relatively high entropy and a disassortative mixing pattern along with very low betweenness centralization. Thus, although the structure was relatively

Table 9.5 Regression model for entropy

	Estimate	Std. error	t-value	p-value	
Intercept	3.408E^{-1}	1.034E^{-1}	3.297	0.00105	
Intercept* I (Week ≥ 7)	-2.377E^{0}	2.872E^{-1}	−8.277	1.34E^{-15}	*
Intercept* I (Week ≥ 10)	2.825E^{0}	2.679E^{-1}	10.542	$< 2\text{E}^{-16}$	*
Intercept* I (Week ≥ 42)	1.559E^{-1}	1.227E^{-2}	12.711	$< 2\text{E}^{-16}$	*
Intercept* I (Week ≥ 92)	-6.323E^{-1}	1.358E^{-2}	−46.574	$< 2\text{E}^{-16}$	*
Intercept* I (Week ≥ 93)	-8.551E^{-2}	3.586E^{-3}	−23.843	$< 2\text{E}^{-16}$	*
Intercept* I (Week ≥ 204)	-1.623E^{0}	1.652E^{-1}	−9.822	$< 2\text{E}^{-16}$	*
Intercept* I (Week ≥ 250)	3.325E^{0}	1.753E^{-1}	18.974	$< 2\text{E}^{-16}$	*
Intercept* I (Week ≥ 335)	-8.865E^{-1}	6.376E^{-2}	−13.905	$< 2\text{E}^{-16}$	*
Week	2.167E^{-1}	4.226E^{-2}	5.127	4.31E^{-7}	*
Week* I (Week ≥ 7)	4.734E^{-1}	7.967E^{-2}	5.942	5.52E^{-9}	*
Week* I (Week ≥ 10)	-6.932E^{-1}	6.754E^{-2}	−10.263	$< 2\text{E}^{-16}$	*
Week* I (Week ≥ 42)	-1.301E^{-3}	5.413E^{-4}	−2.403	0.01664	
Week* I (Week ≥ 92)	7.719E^{-3}	3.654E^{-4}	21.128	$< 2\text{E}^{-16}$	*
Week* I (Week ≥ 204)	1.489E^{-2}	1.465E^{-3}	10.167	$< 2\text{E}^{-16}$	*
Week* I (Week ≥ 250)	-2.631E^{-2}	1.516E^{-3}	−17.354	$< 2\text{E}^{-16}$	*
Week* I (Week ≥ 335)	6.247E^{-3}	4.225E^{-4}	14.784	$< 2\text{E}^{-16}$	*
Week^2	-2.585E^{-2}	4.219E^{-3}	−6.127	1.91E^{-9}	*
Week^2* I (Week ≥ 7)	-1.718E^{-2}	5.966E^{-3}	−10.263	0.00417	
Week^2* I (Week ≥ 10)	4.310E^{-2}	4.219E^{-3}	10.217	$< 2\text{E}^{-16}$	*
Week^2* I (Week ≥ 42)	-5.709E^{-5}	8.415E^{-6}	−6.784	3.55E^{-11}	*
Week^2* I (Week ≥ 92)	-2.550E^{-5}	2.641E^{-6}	−9.656	$< 2\text{E}^{-16}$	*
Week^2* I (Week ≥ 204)	-3.408E^{-5}	3.243E^{-6}	−10.506	$< 2\text{E}^{-16}$	*
Week^2* I (Week ≥ 250)	5.204E^{-5}	3.298E^{-6}	15.780	$< 2\text{E}^{-16}$	*
Week^2* I (Week ≥ 335)	-1.075E^{-5}	7.085E^{-7}	−15.172	$< 2\text{E}^{-16}$	*

*Denotes statistically significant variables ($\alpha = 0.05$)

egalitarian, there was a tendency for some individuals to connect and coordinate with other members who were different from themselves, maximizing diversity as a result. The high values for outbound degree centralization also suggested that some individuals were much more extroverted than others in seeking collaborative partners. Overall, the configuration resembled that of a simple structure, dominated more in spirit than in fact by a very hard working, minuscule leadership cadre.

As time passed, however, betweenness centralization increased, and entropy at least momentarily did so as well. If we also factor in the relative decline in disassortativity, we can say that Wikipedia was gradually shifting toward a machine bureaucracy, in which some members played an important role as power brokers and those elites also grew to interact with one another more often based on the decline in disassortativity.

Between weeks 54 and 92, both entropy and betweenness centralization declined. This indicates a period of reorganization during which some leaders took charge by working hard and focusing on particular parts of the project rather than by scattering

Breakpoint/Period	Social Entropy	Assortativity	Betweenness Centralization	Outbound Deg. Centralization	Inbound Deg. Centralization	Configuration
Initial Values	Elevated	Negative	Very Low	Very High	Somewhat High	Simple Structure
Weeks 4-6	↓	↓	↓	↓	⇨	
Week 7	Elevated	Negative	Very Low	Very High	Somewhat High	
Weeks 7-9	↑	↓	↓	↓	⇨	→ Machine Bureaucracy
Week 10	Very High	Negative	Very Low	Very High	Somewhat High	
Weeks 10-41	↑	↓	↑	↓	⇨	
Week 42	Very High	Negative	Moderate	Elevated	Somewhat High	
Weeks 42-53	↓	↓	↑	↓	⇨	
Week 54	Elevated	Negative	Low	Moderate	Somewhat High	→ Transition
Weeks 54-91	↓	↑	↓	↓	⇨	
Week 92	Moderate		Very Low	Low	Somewhat High	
Week 93	Moderate	Low dissassortative	Very Low	Low	Somewhat High	
Weeks 93-141	↑	↑	↓	↓	⇨	
Week 142	Elevated	Low dissassortative	Very Low	Elevated	Somewhat High	
Weeks 142-203	↑	↑	↓	⤴	⇨	
Week 204	Elevated	Low disassortative	Very Low	Very High	Somewhat High	
Weeks 204-206	↑	↑	↓	↑	⇨	→ Adhocracy
Week 207	Elevated	Low disassortative	Very Low	Low	Somewhat High	
Weeks 207-249	⤴	↑	📋↓	↑	⇨	
Week 250	Elevated	Near Zero	Very Low	Low	Somewhat High	
Weeks 250-334	↓	↑	↓	↑	⇨	
Week 335	Elevated	Near Zero	Very Low	Low	Somewhat High	
Weeks 335-491	⬌	⬌	⬌	↓	⇨	
Ending Values	Elevated	Near Zero	Very Low	Low	Somewhat High	

Fig. 9.4 Configurational changes over time

their efforts across domains. In this respect, they grew increasingly dedicated and bound to particular parts of the work. Their role as "makers" increased, while that of "coordinators" decreased. Overall, these trends signaled the beginning of a diffusion process in which a functional leadership of "leading by doing" emerged. This was a transition period, during which time the system essentially sought a new structure after having shifted off its initial trajectory toward the machine bureaucracy archetype.

The role of the leaders was consolidated after week 92. During this last major interval, although entropy fluctuated along the expected paths of growth and maturation, it eventually found its "steady state," especially after week 250. Additionally, and most importantly, assortativity almost reached zero after week 250, while two forms of centrality were consistently low (betweenness) or very low (outbound degree). All three were also very stable after week 92, with minimal erratic vacillations in observed values. Due to these "leveling off" processes, we can broadly say that Wikipedia shifted to an adhocratic state during this phase.

After week 92, entropy increased for a while, indicating an inflow of new collaborators, only to peak around week 250 and to level off after week 335. Also, betweenness centrality and assortativity, two central measures of more formal organizational configurations—especially those of a bureaucratic sort—tended toward zero. The only type of centralization that saw a recovery was inbound degree centralization, which suggests that some contributors drew others' efforts toward them, but the lack of a similar growth in outbound degree centralization indicates that there was not an especially active core group that actively sought out connections with others.

Overall, when we consider all five dimensions that may be used to characterize organizational configurations, we can say that Wikipedia gradually moved from a simple organization (at the beginning of the data set) toward a machine bureaucracy (until week 54), entered a period of flux between weeks 54 and 92, and finally developed into an increasingly accentuated adhocracy (from week 93 onward).

9.7   The Path Toward Adhocracy

The division into phases articulated here only partially overlaps with the phases of evolutionary development described in Chap. 5. For instance, in Chap. 5 we discussed four phases, while here we have only three major ones if we exclude the period of transition from weeks 54 to 92. There are two reasons for this difference. First, in Chap. 5, we took a strict and unidimensional view of evolution that was derived from the community of practice framework. In that chapter, we followed the evolution of the practice of editorial collaboration, specifically aiming to identify the expected phases of incubation, coalescence, growth, and maturity.

Further, only one aspect of the organization's structuration was observed for the identification effort in Part I of this volume. The description was, at the same time, formal and meta-theoretical, and the domain-specific nature of the organization and the substance of any given change were set aside. In Part II, especially in Chaps. 8 and 9, we added a substantive investigation of the transformation process. To put it another way, in Chap. 5 we looked at "how" organizational change took place, while in the chapters since, we have considered "what changed and to what effect."

This brings us to the second reason for the simpler, tripartite division of Wikipedia's organizational evolution, as proposed in this chapter. When all is said and done, as we first suggested in Chap. 1, Wikipedia's evolution has a beginning, middle, and end. Both transformation models—the one presented in Chap. 5 and the one detailed in this chapter—can be seen through this lens. The ways in which the evolutionary model presented in Chap. 5 complies with these three stages are self-evident. As for this change model presented in this chapter, mapping it onto a beginning, middle, and end paradigm is also relatively easy. Wikipedia was born as a simple, mom-and-pop organization. As it matured, much like many traditional corporations, it grew into something resembling a machine bureaucracy. In the long run, however, its configuration eventually gravitated toward that of an adhocracy.

More to the point, in the beginning, Wikipedia was driven by a group of dedicated individuals who worked in an entrepreneurial context. For a period, covering a bit less than the first 2 years of growth, the organization found a viable manner of functioning within a quasi-bureaucratic structure that included implicit task leaders and coordinators. In the end, however, its explosive success was only possible because it accepted and embraced an adhocratic regime. Although the adhocratic form of Wikipedia that we see today has a leadership structure, those leaders are simply those who do the most work, not those who give orders, arbitrate disputes, or otherwise make guiding decisions for the organization. In this adhocratic phase, Wikipedia's social structure could be likened to that of a gigantic marathon. The leaders, then, are those who run the fastest while maintaining the highest level of stamina. The pace setters determine the rhythm and the direction of the entire project. In other words, those who are most dedicated, as exhibited through their work alone, enjoy, as their reward, more of a voice and more control over Wikipedia as a whole.

9.8 Conclusion

As we have now shown, the development of online collaboration spaces, such as that surrounding Wikipedia, does indeed proceed through a series of phases, ultimately culminating as an adhocratic order that combines several interesting collaborative characteristics. Of these, the combination of fluid leadership roles with the domination of the "1%" contributor group is probably the most intriguing, as it provides a tangible representation of the theorized flexibility that an adhocratic structure entails.

The discussion about adhocracy cannot be completed without reiterating the point made earlier about the double meaning of the term. In Part I of this book, adhocracy was introduced as a framework and was considered through the lens of elite turnover. In Part II, we looked at adhocracy as a specific organizational configuration that was defined by five different dimensions. With this redefinition of adhocracy, we transition from simpler questions about "how" to more sophisticated inquiries about "what" and "to what effect."

Of course, the impact of this analysis is not merely to capture a few changes in variables, nor only to specify the meaning of adhocracy. Rather, this also serves as a first step toward validating the entire synthesized theoretical model articulated in the previous three chapters. Focusing first on the distinction between evolutionary trends and revolutionary moments allows us to quantitatively determine the exact phases through which Wikipedia progressed in its development. Using those same factors, we can also identify the trajectory taken by Wikipedia during its growth. And, as will be shown in Chap. 10, we also relate those key changes to some concurrent socio-technical organizational events, providing a better understanding of Wikipedia itself, yielding insights about the needs of online collaborative organizations in general, and solidifying the synthesized theoretical model provided in this volume for future use across online and offline contexts.

References

Britt BC (2013) Evolution and revolution of organizational configurations on Wikipedia: a longitudinal network analysis. Dissertation, Purdue University

Britt BC (2015) Stepwise segmented regression analysis: an iterative statistical algorithm to detect and quantify evolutionary and revolutionary transformations in longitudinal data. In: Matei SA, Russell MG, Bertino E (eds) Transparency in social media: tools, methods and algorithms for mediating online interactions. Springer, New York, pp 125–144

Freeman LC (1979) Centrality in networks: I. Conceptual clarification. Soc Networks 1:215–239

Mintzberg H (1979) The structuring of organizations: a synthesis of the research. Prentice Hall, Englewood Cliffs

Wagner AK, Soumerai SB, Zhang F, Ross-Degnan D (2002) Segmented regression analysis of interrupted time series in medication use research. J Clin Pharm Ther 27(4):299–309

Zeileis A, Kleiber C, Krämer W, Hornik K (2003) Testing and dating of structural changes in practice. Comput Stat Data An 44:109–123

Zeileis A, Shah A, Patnaik I (2010) Testing, monitoring, and dating structural changes in exchange rate regimes. Comput Stat Data An 54:1696–1706

Chapter 10
Breakpoints and Concurrent Factors

10.1 Introduction

Up to this point, we have articulated a synthesized model of organizational change, and we subsequently detailed some initial findings about the moments at which the organizational configuration of Wikipedia—or the manner in which that configuration was changing over time—shifted.

Our perspective has, so far, been social structural, as we observed only the collaborative factors that were associated with Wikipedia's organizational changes over time. However, this leaves out the forces that drove such changes. These are the factors that operate at the level of organizational policy, deliberate and unconscious organizational engineering and management, and socio-technical infrastructures. These concurrent factors further coincide and have points of convergence with the changes we detected in Wikipedia's structural evolution, suggesting that they may play a casual role in the revolutions observed on Wikipedia.

To identify candidates for concurrent, potentially causal factors, we conducted an archival analysis of major events that occurred within the Wikipedia community, its policymaking environment, and its socio-technical design processes. In this analysis, we looked at the revisions made to key Wikipedia policy pages as well as contributions made to their associated "Talk" pages, in which editors discuss the revisions being made, in addition to larger societal events that may have affected Wikipedia. The findings of the archival search were used to assess the significant internal and external dynamics acting upon the Wikipedia collaborative community.

For the purposes of this chapter, we will focus only on the internal policies under study; the full analysis including external phenomena is detailed at length in Appendix B. With that in mind, our investigation included all of the pages listed under the "Principles," "Content standards," and "Working with others" categories on Wikipedia's "Policies and guidelines" page (Wikipedia 2016b, d). The "History" page for Wikipedia (2017a) was also observed for the sake of capturing other key

© Springer International Publishing AG 2017   143
S.A. Matei, B.C. Britt, *Structural Differentiation in Social Media*, Lecture Notes in Social Networks, DOI 10.1007/978-3-319-64425-7_10

moments in the community's history beyond policy changes. Major concurrent factors were further explored using discussion forums frequented by longtime Wikipedia users (such as Wikipedia-L, archived at https://lists.wikimedia.org/pipermail/wikipedia-l) as well as the growing literature dedicated to Wikipedia as a community project (Jullien 2012; Tkacz 2014).

In this enterprise we seek central organizational events and aim to present the change process in terms that are as detailed and tangible as possible. In this, we take an idiographic approach (Lamiell 2014). As a term whose origins lie in the history of ideas, idiography is the practice of identifying unique events, decisions, motivations, or behaviors that occur above and beyond what regular trends might predict (Luthans and Davis 1981). We adopted this approach to account for unique events, which can act as "switches" in the otherwise incremental development of Wikipedia. The main justification of this approach is to avoid the temptation of overemphasizing and rationalizing the evolutionary process. Living organizations, like Wikipedia, are not absolutely preordained to take a very specific path. While they can and do follow some regularities that are imposed by various constraints (material, formal, efficient, or teleologic, to use the old Aristotelian adage), they are not entirely predetermined, in much the same way that the growth of an organism is only partially predicted by genes and biology. Accidents, context, and human decisions may have a significant influence on collaborative evolution and its organizational outcomes.

In addition, we will examine the manner in which those factors shifted Wikipedia's trajectory of growth, building on the theory of organizational motors that was addressed in Chap. 8. To recount, there are four modalities of change: life-cycle, evolutionary, dialectical, and teleological. Life-cycle motors are those internal forces that promote the organization's natural, uninhibited life span, much like the way in which a child will grow taller over time in the absence of other prohibitive forces to prevent it, due solely to his or her own life cycle. In the context of organizations, life-cycle motors effectively coax the organization along a predetermined linear sequence that leads toward an inevitable conclusion, with little more flexibility than a railroad car moving down a set of train tracks.

Evolutionary change, in contrast, is driven by the environmental forces that promote and maintain the three key evolutionary processes: variation, selection, and retention. For both life-cycle and evolutionary motors, the organization has little power to resist the change in question (at least, if its members want the organization to survive). The major difference between the two is whether the change involves the organization following its own natural trajectory or whether the organization is instead driven toward a new form or process by environmental conditions or constraints.

Dialectical change emerges through conflicts between actors who hold competing goals or values. The resolution of those conflicts through dialogue may establish new norms within the organization and throughout the larger environment in which the organization is embedded. And lastly, teleological change unfolds when the leadership unit within the organization pushes it forward in pursuit of a desired "final state," without necessarily engaging in extensive dialogue and negotiation about the change in question.

Overall, the efforts described in this chapter were undertaken to identify the major policy, technology, and community events occurring immediately before, during, and after each breakpoint. This deeper assessment of concurrent factors allows us to examine the extent to which the observed configurational changes were impacted by particular regulatory, managerial, technological, or social factors. The goal in doing so is to link the observed internal configurational changes with the various motors that contributed to those changes, firmly unifying the work of Mintzberg (1979, 1989) with that of Van de Ven and Poole (1995).

10.2   Identifying Concurrent Factors and Change Modalities

The analysis in this chapter considers all major breakpoints identified through stepwise segmented regression (Tables 9.1, 9.2, 9.3, 9.4, and 9.5) in all five dimensions, as shown in Figs. 10.1 and 10.2. These correspond to weeks 7, 42, 54, 92, 142, 204, 250, and 335.[1] We examined the various major events that occurred in the 4 weeks prior to each breakpoint as well as the week afterward in order to properly account for events that may have had a delayed effect or that may not have been widely recognized until after their impact.

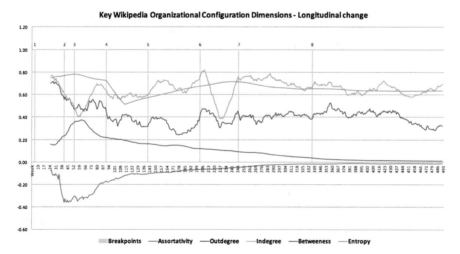

Fig. 10.1 Organizational change dimensions over time. This figure uses moving average values to smoothen periodic changes for the sake of interpretability. Moving averages calculated with a 24-week lag do not include the breakpoint occurring at week 7. Complete data can be found in Appendix A, Table A.2

---

[1] For the purposes of this assessment, multiple breakpoints occurring within 5 weeks of one another were considered to constitute a single breakpoint, as it would otherwise be impossible to properly distinguish the motors that affected different breakpoints in such close temporal proximity.

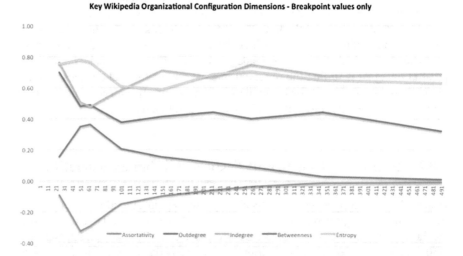

Fig. 10.2 A simplified version of Fig. 10.1 that includes only the values at each breakpoint, which are then connected by straight trend lines. This figure shows how values moved linearly from breakpoint to breakpoint. Vertices in the lines indicate breakpoints

We will first examine the events surrounding each breakpoint sequentially. After this "blow-by-blow" account, we will explore the full set of breakpoints in the context of their respective idiographic and concurrent organizational change factors. This allows for the assessment of momentary forces at work as well as any consistent themes that pervaded the full data set. In the process, we will also discuss the modality of the change process at each given point in time when appropriate.

10.2.1   First Breakpoint: Week 7

During its first weeks of life, in January–March 2001, Wikipedia was a tiny organization animated by the ideals of the "open source" software movement and run by a very small group of people. Yet, its growth was promising, as indicated by Appendix A, Table A.1. Twelve authors already generated 648 articles. At the same time, the work was rather dispersed: only 905 revisions, or about 1.4 per article had been made. Yet, if we instead divide the number of revisions by the number of authors, the work is rather significant, with approximately 75 revisions per author. Entropy was relatively high during this period, reaching almost 0.7 on a scale from 0 to 1. So, although the load was carried by a small group of contributors, the effort was rather evenly spread among these individuals.

At this stage, policies and rules were more implicit than explicit. In fact, the "Policies and guidelines" page used as an index of major Wikipedia policies did not yet exist, as the community was still too small for strict rules to be necessary.

Considering that a small group had, to that point, produced relatively minimal content, there was hardly a need for a rigid set of guidelines dictating interpersonal behavior and contribution quality.

However, although there was hardly an extensive system of Wikipedia rules at this time, Jimmy Wales and Larry Sanger did formally establish one of the organization's most important policies, the "neutral point of view" standard for revisions (Wales 2001; Wikipedia 2017a, c), around this time. This policy, which was first posted to Wikipedia on February 16, would eventually become one of Wikipedia's three "core content policies" that govern all information added to the encyclopedia.

10.2.2   Second Breakpoint: Week 42

The shift in collaborative dynamics captured after this date by our metrics, especially entropy, appeared to reflect a major change in Wikipedia's social infrastructure. In this and the weeks that followed, the entropy of collaborative effort on Wikipedia reached its highest level, at almost 0.8. At the same time, on November 1, 2001, the third day of week 42, Wikipedia cofounder Larry Sanger created the "Policies and guidelines" page that would eventually serve as the foundation for the community's array of policies (Wikipedia 2017d). The creation of this article in particular began the process of formalizing Wikipedia policies in a manner that offered a degree of permanence as a master code of conduct, much more so than the earlier drafts of individual policies like Wikipedia's "neutral point of view" (see Wales 2001), and it ultimately served as an important step toward guiding editors' efforts in a positive, productive direction.

Although entropy continued to increase in the weeks immediately following this change, which might suggest a progression toward relative egalitarianism, in the long run Wikipedia's policies throttled this movement. The effect of Wikipedia's formalized policies would be felt in the more distant future, when the project became increasingly dominated by some voices who often used the rules to defend their edits (Matei and Dobrescu 2011), while coeditorial connections between high- and low-activity editors concurrently declined in prevalence.

10.2.3   Third Breakpoint: Week 54

In terms of broader organizational dynamics, this was a major organizational configuration shift point, which was first identified in Chap. 9. At this point, entropy and betweenness centralization began declining, indicating a process of reorganization. Some leaders took charge by working in a concerted, focused manner. More broadly, week 54 signaled the beginning of a diffusion process in which a functional leadership style of "leading by doing" emerged. This acted as the beginning of a transition phase between bureaucracy and adhocracy.

In the weeks prior to this breakpoint, Wikipedia's editorial velocity had begun to rapidly increase. Between week 42 and week 54, the number of articles that had been created and revised jumped from roughly 13,000 to 20,000, and the number of authors also climbed from approximately 250 to 500. In other words, Wikipedia almost doubled in size.

More authors and more articles also meant more computing time. The growing server load started to take its toll, causing persistent performance issues through the community. And although Wikipedia was still mostly unnoticed by the outside world, members were already concerned that even a single article in a major media outlet might spur a flood of new members to the encyclopedia, which could derail the site entirely.

In order to cope with this success, university student Magnus Manske began developing a new PHP script in summer 2001 that would better manage Wikipedia's resources. This database-driven application was designed to reduce the performance problems that plagued the previous UseModWiki content management system.

On January 25, 2002, halfway through week 54, Manske's script replaced the UseModWiki engine on Wikipedia. The use of his script would later become known as "Phase II" in contrast with the "Phase I" UseModWiki engine (Mediawiki 2016; Wikipedia 2017a). Its introduction enabled a greater amount of activity on Wikipedia without users enduring the performance issues that had come to plague the UseModWiki-powered Wikipedia.

As a secondary consideration, it should be noted that when the database software was transitioned from Phase I to Phase II, only the most recent revision of each article was converted to the new system. Another leading Wikipedia developer, Brion Vibber, partially restored previous iterations of each article in September 2002, but it does not appear that the full set of these initial revision records was ever completely restored (Mediawiki 2016). The problem of incomplete data was exacerbated by the small sample sizes in the beginning stages of Wikipedia's development. Consider that across Wikipedia's first 53 weeks, revisions were attributed to only 444 registered members, while the significance of those revisions as measured by their combined delta scores was only 5,786,348. In contrast, in week 250 alone over 3,000 new editors joined the Wikipedia community, as the number of registered members who had made at least one revision increased from 120,630 to 123,715. Further, the combined delta scores of all revisions made during week 250 amounted to approximately 15,000,000, so based on the present data set, more than twice as much editing activity occurred during week 250 alone than in Wikipedia's first 53 weeks combined.

When taken together, the small number of Wikipedians during its first year of development and the fact that the available data on editor activity is incomplete may cast some doubt upon the accuracy of measures prior to week 54, as smaller samples are likely to generate much greater week-to-week variance and the incomplete data further has the potential to be systematically biased. As such, any breakpoints that emerged in the year prior to week 54 must be treated with greater scrutiny.

10.2.4   Fourth Breakpoint: Week 92

As can be seen from Fig. 10.1, there was a major drop in entropy at the site of this breakpoint, the greatest and most dramatic in the entire history of Wikipedia, with entropy falling from a relatively high level of 0.7 to just under 0.5. This coincided with the launch of Derek Ramsey's "bot," or automated user account, which rapidly added about 10,000 new articles to Wikipedia corresponding to cities listed in the US census (Wikipedia 2017a), doubling the number of articles on the site (see Appendix A, Table A.1). Although bots had already seen limited use on Wikipedia in handling repetitive tasks, the so-called "rambot" implementation was the first large-scale use of bots to systematically write large amounts of content with minimal human supervision.

Rambot created its first article on October 5 (Wikipedia 2016a), and 2 weeks later it dramatically increased its activity on Wikipedia, generating a deluge of new articles for months to come. Its use would set the standard for the numerous bots created ever since (Wikipedia 2017b). More immediately, rambot transformed Wikipedia's division of labor, taking sole responsibility for almost half of the content added to Wikipedia over a period of several weeks and dramatically shifting the community entropy as a result. The dramatic drop in entropy that can be seen in week 92 is the product of rambot's intervention on the site. It is important to note, however, that in time, bots declined in importance; historically, their overall impact comprises less than 5% of all Wikipedia content (Steiner 2014).

10.2.5   Fifth Breakpoint: Week 142

The week 142 breakpoint fell in September–October 2003, almost exactly 1 year after the week 92 breakpoint. The editorial dynamic, as can be seen from Fig. 10.1 and Appendix A, Table A.2, suggests a slow but significant increase in entropy, and the number of articles and revisions was also about to dramatically increase. This is the phase that we identified in Part I of this volume as one of "growth." We did not, however, discuss it in Chap. 9 since it did not mark sufficient changes in several dimensions. As we will indicate below, this is a momentary signpost that only marks a "waypoint" in Wikipedia's evolutionary process.

Much like our previous breakpoint (week 92), this breakpoint coincided with very few revisions to Wikipedia policies. The most noteworthy activity during this time was an anonymous editor vandalizing the "Policies and guidelines" page on September 8; it was reverted to its original version within 2 min.

10.2.6   Sixth Breakpoint: Week 204

The breakpoint for week 204, which occurred at the end of 2004, represented the end of the growth period (see Fig. 10.1). Although entropy was inching upward (at 0.68) and was near the top of the entropy "bulge" during the second half of Wikipedia's evolution, its values were still not as high as those seen during the initial stages of the project (see Appendix A, Table A.2). More importantly, all of the other organizational change dimensions were becoming less relevant. Wikipedia would soon reach a state of maturity and stability, after which it would remain in a relatively steady state for the rest of the recorded period.

Likewise, in almost 4 years, Wikipedia had accumulated over 700,000 articles revised by over 40,000 members. Clearly, Wikipedia stood as a growing social medium with a significant following. It also started getting substantial attention from the press during this time, appearing in 34 articles in major US newspapers during the 5-week assessment period surrounding this breakpoint (for details about the press coverage, see Appendix B). Not all of the attention was positive, however. As Wikipedia grew, it had attracted the attention of individuals who saw it as the perfect target for vandalism. One newspaper article noted that the pages for George W. Bush and John Kerry had to be locked in the weeks before the 2004 US presidential election, yet on November 2, election day, Bush's page was nonetheless vandalized twice: the first time, a "Kerry for President" banner appeared over a photograph of Bush, and the second time, his picture was replaced with a photograph of Adolf Hitler (Boxer 2004).

In response to the events that brought Wikipedia's credibility into question, a "Verifiability" policy was added to the site's "Policies and guidelines" page (Wikipedia 2017d). This more explicitly highlighted the need for editors to cite external sources when adding new content, further raising the standards for Wikipedia content and the effort required to make an acceptable contribution. "Verifiability" would eventually become one of the most important policies on Wikipedia, as is discussed in more detail below. While the policy did not throttle participation on its own, it would, in time, contribute to stabilizing the participation and contribution regime.

10.2.7   Seventh Breakpoint: Week 250

The week 250 breakpoint, which marked the end of the maturation phase, was found in late September 2005. In the period between week 204 and week 250, entropy reached a local maximum and then started leveling off. Assortativity and betweenness centrality, two core indicators for social aggregation in complex bureaucracies, steadily approached zero, making it clear that Wikipedia was operating as an adhocracy in full. Along with these configurational developments, content production started to grow exponentially. The number of revisions made to Wikipedia would double in the following 6 months and would double again in the 6-month interval after that.

As Wikipedia neared the end of its fifth year, it increasingly became a topic of public debate. It was mentioned in 95 separate articles in major US newspapers during the five-week period corresponding to the week 250 breakpoint (see Appendix B), although again, not all of the attention was positive. On October 3, author Nicholas Carr highlighted significant stylistic problems with Wikipedia on his blog, specifically using the Bill Gates and Jane Fonda articles as examples of "incoherent and dubious" writing (Carr 2005; Langberg 2005). His comment, and the discussion that followed, spurred Wikipedia cofounder Jimmy Wales to send several messages of his own through the Wikipedia WikiEN-l mailing list on October 6 and 7. Although he made it clear that he disagreed with some of Carr's central points, Wales conceded that "[t]he two examples he puts forward are, quite frankly, a horrific embarrassment. [sic] Bill Gates and Jane Fonda are nearly unreadable crap. Why? What can we do about it?" (Wales 2005).

Wales' concession that Wikipedia content may be flawed, at least occasionally, quickly became a hot topic in the press, especially since Wikipedia was itself becoming increasingly legitimate competition for traditional news media. Many in the press latched onto Wales' words, using them to justify the continued existence of journalism professionals like themselves (e.g., Harkin 2005; Orlowski 2005).

Yet, with greater controversy came not only greater scrutiny but also greater popularity. By week 250, Wikipedia stood as a rapidly growing resource for information and learning, and public discourse around it would hardly stifle that growth.

10.2.8 Eighth Breakpoint: Week 335

This final breakpoint at week 335, corresponding to late May 2007, marked the point at which Wikipedia reached a steady state. Setting aside minor vacillations and small periodic shifts, all five dimensions being measured were almost completely level. Even entropy had become linear and flat.

Two major dynamics followed this breakpoint. First, the collaborative dynamic surrounding many Wikipedia policy pages dramatically changed. Policy pages, documents that record the major decisions made by a group of active and knowledgeable users regarding editorial contributions, content, and interaction dynamics, became increasingly difficult to edit. The few revisions made to Wikipedia policies during this time tended to involve mere copy editing or, at most, clarifications of existing policies. They were either actively protected by old-timers, who watched changes and promptly reverted them if they did not fall in line with the established consensus, or they were "padlocked" with a protection mechanism that prevented changes by ordinary users. By early 2006, only activity on the corresponding talk (commentary) pages had increased and only for some of the more controversial policies. Many others had organically reached states of relative stability, as individuals who tried to alter them struggled to obtain a consensus from the community, with many fellow editors seeing little need to overturn long-standing policy.

Second, a new article by Nicholas Carr, the very same author who lambasted Wikipedia just 2 years earlier (Carr 2005, 2007), suggested that Wikipedia had become a favored source of reference materials for search engines. By this point, as Carr suggested, many searches for common nouns returned a Wikipedia article as one of its top results. This was not a concession about Wikipedia's quality, but if nothing else, it illustrated how far Wikipedia had come to be acknowledged, if only based upon its widespread use.

Likewise, Wikipedia's image in the press further evolved. Two hundred one major newspaper articles mentioned Wikipedia in the five-week period surrounding the week 335 breakpoint alone. More importantly, many of the articles mentioning Wikipedia ceased to explain its status as an online encyclopedia that anyone could edit, instead merely using it as a source or to provide examples of the key points being made within the article (e.g., Cohen 2007; Whyte 2007). In other words, Wikipedia's existence no longer needed to be explained—its use had become so widespread that its role in society was common knowledge.

10.3   Systematic Overview of the Factors Shaping Wikipedia's Evolution

Across these eight breakpoints, a consistent theme emerged, with multiple breakpoints aligned with major Wikipedia policy changes. The week 7 breakpoint fell in line with the introduction of the neutral point of view standard, for instance, while the week 42 breakpoint corresponded to the creation of the "Policies and Procedures" page. There were also a few prominent Wikipedia events outside of policy changes, especially the institution of the Phase II software in week 54 and Jimmy Wales' public admission of substantial writing problems with Wikipedia content near the week 250 breakpoint.

Let us start this overview discussion with week 42, which coincides with the creation of the "Policies and procedures" page. This page served as the starting point for formalizing standards for content and behavior on Wikipedia, giving them a degree of permanence by collecting the many distributed policies into a single guide and establishing them as a single, unified code of conduct that stood as the sole authority for proper Wikipedia discussion and editing decorum.

When Larry Sanger organized the array of Wikipedia policies to make them easier for editors to find and follow, his actions further established the policies' importance to the community. This initiative highlighted the clear expectation that editors should know and follow the guidelines. The easier it is to locate and understand the "rules," after all, the less of an excuse there is for not following them (the "Ignore all rules" policy notwithstanding).

Sanger's action stood as a significant teleological motor, to use the terminology proposed in Chap. 8, thus raising the standard for contributions to Wikipedia. Since it became clear to users that they would have to devote a substantial amount of effort in order to make an acceptable revision to a given Wikipedia article, editors were prevented from simply flooding Wikipedia with lackluster content to boost their own standing through the sheer number of contributions. It would be natural to expect entropy to decline as a result of these stricter standards—not all users have the time to develop pages upon pages of well-crafted content, so stratification between top editors and the rest would be quite likely—and values of the measure did indeed fall after Sanger's addition.

Moving to the week 54 breakpoint, the major change in Wikipedia software architecture prompted a key organizational shift. Wikipedia moved from the UseModWiki database engine to the PHP script written by Magnus Manske. On the broadest scale, this had two very important effects on the project and the present data set. First, at least for a time, it resolved the performance issues that had been making editing more difficult for Wikipedians since the middle of 2001. This facilitated much greater, more consistent activity on Wikipedia without generating substantial slowdowns like what some editors had experienced in the past. Just as importantly, however, when Wikipedia changed to the Phase II architecture, not all revisions to existing articles were maintained, and the records of earlier revisions were only partially restored later.

In fact, this revelation about the incomplete data prior to week 54 might place the week 54 breakpoint itself in doubt. If there was a systematic bias in values of the measures taken before week 54, then that could, conceivably, be the cause of the breakpoint itself. On the other hand, if revisions were lost at random rather than as the result of a particular pattern, then any effect on values of the five key measures used in this study would be negligible, so the breakpoint would indeed have been the result of behavioral changes among editors rather than a mere result of missing data.

With that in mind, it is to be expected that the measures taken before week 54 would have greater variance than those that came afterward due to the reduced sample size. This alone would not have erroneously caused the week 54 breakpoint, however, as changes in variance would not have presented as breakpoints within the regression-based approach for identifying phase shifts.

The only way, then, that the missing data would have falsely generated the week 54 breakpoint would be if it presented a systematic bias that altered a given measure to make it consistently higher or lower than its true value. In that instance, we would expect the week 54 breakpoint to be punctuated by a substantial discontinuous jump like that observed for entropy at weeks 92 and 93 (see Fig. 10.3), not merely a change in the slope of the line. If, for instance, the measures for betweenness centralization were consistently much greater than they should have been, that shift would be immediately reflected at week 54 as the bias was instantaneously corrected. A close examination of Fig. 10.4, however, reveals that this was not the case.

Certainly, the fitted values for betweenness centralization changed dramatically at week 54, but the observed values themselves began a slower path downward

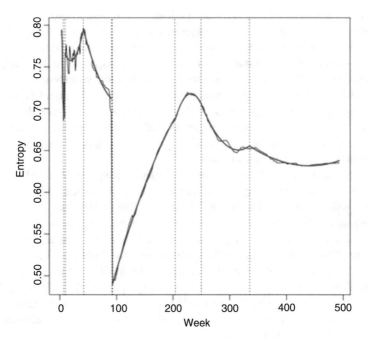

Fig. 10.3 Plot of entropy raw data and its fitted values from the regression model, with breakpoint locations marked with *dotted lines*

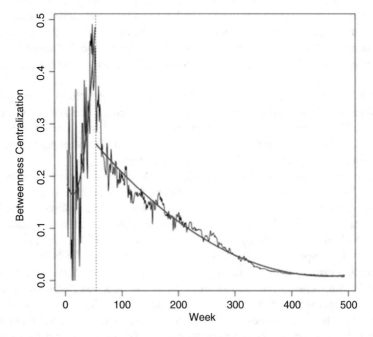

Fig. 10.4 Plot of betweenness centralization raw data and its fitted values from the regression model, with the breakpoint location marked with a *dotted line*

rather than instantly changing to match the rightmost portion of the regression line. Since betweenness centralization began a sharp decline at week 54—not an immediate reversion back to "true" values—it suggests that there was no systematic error in the measure prior to week 54. Rather, the community merely changed course as a result of the transition to the Phase II architecture, allowing members to more freely cross boundaries and connect with one another, escaping the bureaucratic structure that had been developing over Wikipedia's first year.

To return to the discussion of the week 54 breakpoint, as Wikipedia grew, it was inevitable that it would eventually have to change its software in order to allow more editors to offer more contributions without the system breaking down. A growing community requires a larger infrastructure, after all. As such, this upgrade to the Phase II architecture and the resulting shift in Wikipedia's configurational trajectory were the result of a life-cycle motor, to use the terms presented in Chap. 8. To put it simply, just as a growing child will need larger clothes over time, by the end of its first year the community of Wikipedia editors had outgrown its initial system.

Just as the continuous nature of the week 54 breakpoint helps us to verify its veracity, we may also trust in the continuous breakpoint observed at week 42. At this point, entropy clearly peaked rather than exhibiting a discontinuous shift that would have indicated a systematic bias for measures taken prior to week 42. Statistically speaking, the fact that this breakpoint also appears in the form of continuous connections between the line segments on either side suggests that it was not merely spawned by systematically biased observations on one side of the breakpoint, despite the greater variance for observations prior to week 54.

With that being said, the lost data prior to week 54 may weaken the value of the breakpoint for week 7. As is clear from Fig. 10.3, the first two segments of the plot (corresponding to weeks 4–6 and 7–9, respectively) are nearly vertical, suggesting dramatic changes even within these segments. If we were to accept this breakpoint as it is, we would have to conclude that entropy started near its maximum in week 4 and then plummeted, yet suddenly reversed direction and skyrocketed beginning in week 7.

While this is within the realm of possibility, the following might be a better explanation. Consider, instead, the aforementioned fact that in Wikipedia's first 10 weeks, very few people on earth had ever heard of it, so there were exceedingly few individuals taking part in what was, at the time, merely an experiment designed to attract more attention to a related effort, Nupedia (Mediawiki 2016). By week 10, only 21 unique editors existed within the Wikipedia data set, and not all of their contributions may even have been available in full due to the revisions lost during the change to Phase II. With such little activity in the community, there was a great deal of week-to-week variability in the community based on how a very small number of people happened to behave on any given day. In this light, the dramatic changes that occurred in Wikipedia's first 10 weeks seem more like random fluctuations than fundamental changes to the organizational configuration.

This leaves the introduction of the neutral point of view policy as the only other realistic explanatory mechanism for the breakpoint observed at week 7. It is reasonable to suspect that the introduction of this critical policy might have had some effect. But at the same time, this foundational policy was introduced on February 16, 2001, barely a month after Wikipedia's January 15 launch, making it difficult to separate the policy from the launch itself. After all, Wales himself framed the policy as a mere clarification of Wikipedia's role as an encyclopedia, something which was implicit in the creation of Wikipedia itself. Certainly, the formalization of Wikipedia policies can have a significant effect, as has been previously discussed. Yet, since this initial policy came so soon after Wikipedia's launch, at a point in time when the entire data set still contained just eight registered members, it is easy to treat the policy invocation as part and parcel with the launch itself.

More to the point, even if one believes that the neutral point of view policy really could offer an explanatory mechanism for a breakpoint, the fact remains that the breakpoint itself is in doubt. With such high variability within the first few weeks of the data set due to both low membership and incomplete revision data for the very small number of editors within that period, it is much easier to conclude that the breakpoint for weeks 7 and 10 was simply a product of random noise. Yes, it is likely that the neutral point of view policy had some sort of effect on Wikipedia, but a causal link cannot reasonably be established between the creation of that policy and the particular changes observed at weeks 7 and 10. It is much more likely that this breakpoint was nothing more than a statistical artifact borne of missing and minimal data. As already mentioned in Chap. 5, it was not directly used in our first evolutionary model, as it offers minimal explanatory power.

The week 92 breakpoint, in contrast, occurred well after Wikipedia was established, and it aligns with a major concurrent factor. During this period, we saw the activation of "rambot" to expediently create thousands of articles on various US cities using census data. While rambot's efforts began on October 5, 2002, in Wikipedia's 90th week, it was fully unleashed 2 weeks later, during week 92. Within this week alone, the combined delta value of all of rambot's revisions was over 8.86 million, and in week 93, its efforts as measured by delta amounted to approximately 15.1 million. To put that figure in perspective, the combined efforts of all revisions made to Wikipedia by registered users in week 91 summed to a delta value of only 18.6 million, so rambot's presence alone nearly doubled the amount of content being produced on Wikipedia on a weekly basis.

This naturally caused the entropy measure to plummet, as this one automated editor controlled a tremendous amount of Wikipedia's content. However, since few other editors were actively engaging with these newly created articles, many of which dealt with relatively obscure cities, rambot did not become a central figure in the collaboration network, so its presence had minimal effect on the centralization and assortativity measures even as it fundamentally altered the division of labor on Wikipedia. Moreover, in the years to come, its contributions would become a metaphorical drop in the ocean.

Yet, Derek Ramsey's decision to implement rambot constituted a powerful teleological force on the organization, as his actions changed, at least for a while, the division of labor on Wikipedia. With that said, it did not necessarily alter Wikipedia's progress toward an adhocracy as observed through Fig. 10.1. Rather, the trajectory of Wikipedia's development was moving toward an adhocracy before the week 92 breakpoint, and it continued to do so afterward. The only difference is that labor gradually became more evenly divided after the breakpoint (corresponding to rising entropy values) as opposed to the declining entropy levels that preceded the breakpoint. Both trends were, in fact, converging toward the same moderately high entropy values. From a practical standpoint, after the introduction of rambot, more and more members had to arrive and increase their contribution rates in order to counterbalance the significant efforts of the new bot in pursuit of the archetypal adhocracy.

Moving forward, when examining the week 142 and week 204 breakpoints, we may observe that outbound degree centralization—the measure in which the breakpoints were observed—appears to hold a wave shape (Fig. 10.5).

Visually, it does appear that there is a slight dip in the data between weeks 142 and 204, suggesting that the detection of two breakpoints in this area is well justified. Yet, when we consider the entire picture and the significance of key events during this period, these breakpoints seem to be purely developmental, corresponding only to the natural growth (i.e., the life cycle) of the organization. As has already been mentioned, the only significant incident occurring around week 142 was the vandalization of the "Policies and guidelines" page, which was reverted to its original version within 2 min of the incident, thereby mitigating any long-term ramifications. It is thus safe to conclude that this breakpoint is relatively inconsequential as far as revolutionary change is concerned.

The week 204, however, coincided with a more consequential event, as the "Verifiability" policy was added to the master "Policies and guidelines" page around this time. But this development demands closer attention. It is true that "Verifiability" stands as one of the three core content policies on Wikipedia, alongside "Neutral point of view" and "No original research." At the time, however, this trio had not yet been established as the core of Wikipedia's governing policies, so "Verifiability" was just one of many, many policies active on Wikipedia. In fact, when it was added to the "Policies and guidelines" page, it was simply inserted into a block of listed rules, with nothing indicating that it was any more or less important than any other policy in the community. Following this logic, if the mere addition of the "Verifiability" policy to this master list was enough to spark an organizational shift on Wikipedia, then the addition of any other policy should have had a similar effect. Such changes were not observed every time a new policy was added to the "Policies and guidelines" page, so there is no particular reason why "Verifiability," which was not yet deemed especially important, would have been any different.

In the end, if we look at the overall graph outbound degree centralization values, we can observe an apparent sinusoidal curve following the red trend line both before the week 142 breakpoint and after that for week 204. Although the period from weeks 142 to 204 does appear to follow a slightly different path, the overall curve seems to have been unaffected by the segment between these two breakpoints. Therefore, we may tentatively say that the week 142 and week 204 breakpoints are

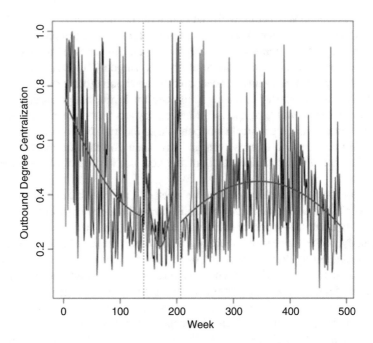

Fig. 10.5 Plot of outbound degree centralization raw data and its fitted values from the regression model, with breakpoint locations marked with *dotted lines*

relatively weak compared to the others that we have discussed, as they do not correspond to any major changes in policies, standard practices, or technological changes, and certainly not organizational configurations; in fact, any effect that they may have held seems to have been temporary. The most that we can claim is the possible role of life-cycle change motors near the end of Wikipedia's third and fourth years, around the time that Wikipedia entered and consolidated its maturity phase. Otherwise, the impact of those motors is limited at best.

The concurrent motors surrounding the week 250 breakpoint, on the other hand, demand closer inspection. At this point, Nicholas Carr's blog (Carr 2005) had drawn Jimmy Wales into a debate on the value of the content produced on Wikipedia, and Wales admitted that Wikipedia content was, in some cases, "unreadable crap." This concession reverberated through Wikipedia, resulting in a wave of activism that led to new tools and methods to improve content quality. For instance, the "Barnstar" program (Wikipedia 2005a), through which active and "quality" users are rewarded with a star of recognition, had appeared in late 2004; in response to Wales' comments, it markedly increased in use. Likewise, the "featured article" designation, which had begun a few months beforehand, similarly increased in prevalence (Wikipedia 2005b). Such quality checks, in concert with the general debate about Wikipedia as a whole, may have attracted new contributors to the project.

In terms of inner collaborative dynamics, values of the entropy measure climbed for quite some time before the week 250 breakpoint, and they had only tailed downward at the end of a curvilinear segment shortly before the breakpoint. After Wales' admission, entropy continued to decline, indicating a growing tendency toward structuration, but it would begin to level off in the months and years to come.

One may surmise that Wales' admission shamed those who read it, leading to a more "take charge" attitude. Of course, the most likely readers of the Wikipedia WikiEN-1 listserv were those editors who were dedicated enough to Wikipedia to subscribe to it. Therefore, it is quite likely that when those dedicated editors read Wales' message, it spurred them into action, and for much of the following year, they took on a greater amount of the editing workload themselves in order to personally ensure its quality. The goal of Wales' message, after all, was to fix the problems that Nicholas Carr had revealed, so it is hardly a stretch to suppose that Wales accomplished his goal through his ensuing conversations with the most dedicated Wikipedia editors. It would make sense, then, that the dialogue gave those editors a reason to work harder to refine the project and to push the community toward higher-quality work overall, so Wales' admission and the discussion about quality control that followed constituted a dialectic change on Wikipedia. In short, the most active editors took matters into their own hands by doing more of the work themselves, which increased the proportion of effort concentrated at the "top" of the project.

This leaves only the week 335 breakpoint. At this point in time, entropy exhibited a minor peak, but it was still almost unchanged from its existing trajectory, and it would remain around the same level for some time to come. As we can see from Fig. 10.1, assortativity was exceedingly close to zero, as was betweenness centralization. Inbound and outbound degree centralization were declining somewhat, but their movement was minimal and generally flat. Thus, this breakpoint serves as an indication that although Wikipedia's entropy level had "leveled off," its stability was driven by more intense lateral connections, which strengthens the claim that it had entered a period of adhocracy.

There were no major changes in the policy and "political" life of the project around this time. The talk pages corresponding to many policies featured a considerable amount of activity, but this had been the case for over a year, so its continuation would constitute an evolutionary change, if anything, rather than a revolution. Similarly, Wikipedia had become common enough knowledge that the news media did not have to explain it, but the project's gradual exposure to the public was also an evolutionary change spanning many years; it had nothing to do with a Wikipedia revolution.

All told, this breakpoint seemed to constitute a fairly minor adjustment of Wikipedia's existing trajectory toward the pure adhocracy archetype. The likeliest conclusion is that the motor behind that adjustment was a purely endogenous lifecycle modality, but even this cannot be claimed with any real certainty.

10.4 Conclusion

This chapter aimed to identify the policy, managerial, and socio-technical factors that influenced Wikipedia's evolutionary process. We observe these forces as unique events that operated alongside internal, structural factors. Consequently, our above discussion was descriptive, aiming to "connect the dots" related to the observed statistical changes in quantitative measurements. To that end, this discussion comes as a complement, not a supplement, to the more rigorous discussion about Wikipedia's organizational change discussed in the previous chapter. When we connect the rigorous, formal model developed in Chaps. 6, 7, 8, and 9—dealing with the structural differentiation factors and organizational configuration dimensions through which Wikipedia's growth is broken into phases—with the more idiographic, descriptive analysis contained in this chapter, we obtain the summary presented in Table 10.1.

For each major breakpoint, we have a major change in policies or internal dynamics. More specifically, the week 42 breakpoint represented a legitimate revolutionary change, as Larry Sanger created the "Policies and guidelines" page as a master code of conduct for Wikipedia editors. Based on the descriptions provided in the previous chapters, this served as a "teleological" motor through which the work started to be more projective and driven by a clearer end goal. This raised the standard for editing, and since more effort was demanded for each contribution, it consequently became harder for a large group of editors to all contribute massive amounts of content. Rather, a few especially dedicated editors provided especially substantial work, while others only contributed to the extent that they were able. In other words, the knowledge-production process became more stratified.

In week 54, Wikipedia's database system was upgraded in order to handle increasing amounts of activity. This resolved long-standing performance issues with the site, allowing users to more freely and quickly explore different articles and collaborate with a wider range of peers on a wider range of articles. We may speculate that this broke up the hierarchy that had been developing between the majority of users who could only spare the time to work on a few isolated articles and those few who weathered the frustration of slowdown and database errors to oversee a wide range of articles—with a better database system, all editors could more quickly engage with many more articles than they ever had before, so they were free to explore many different domains. This was a necessary step along Wikipedia's life cycle, and it allowed the organization to abandon its progression toward a machine hierarchy and begin moving toward an adhocratic structure instead.

At week 92, the automated user later known as rambot began adding countless articles to Wikipedia, quickly becoming responsible for a disproportionate amount of the content produced. This would radically shift the overall division of labor, instantly reducing entropy throughout the community. This was, however, a temporary disturbance; by the end of the 9.5-year period examined here, bots were only responsible for a small minority of content on Wikipedia. And, even in the aftermath of this breakpoint, Wikipedia still progressed toward an adhocracy, with entropy

Table 10.1 Summary of organizational change

	Concurrent factors	Config. change	Measure change	Modality of change
Week 7	Neutral point of view policy created; data unstable, small sample size	→ Machine bureaucracy	Entropy vacillation	
Week 42	"Policies and procedures" page created by Larry Sanger	→ Machine bureaucracy (continuing)	Entropy peak	Teleological
Week 54	Phase II Wikipedia software architecture introduced	Transition	Betweenness peak	Life cycle
Week 92	Widespread use of bots spurred by Derek Ramsey, "rambot"	→ Adhocracy	Entropy drop	Teleological
Week 142	Beginning of maturity phase	→ Adhocracy (continuing)	Outbound degree jump	Life cycle
Week 204	Consolidation of maturity	→ Adhocracy (continuing)	Outbound degree drop, entropy bulge building	Life cycle
Week 250	Jimmy Wales' "unreadable crap" admission, prominent editors calling for content improvements	→ Adhocracy (continuing)	Entropy bulge end	Dialectical
Week 335	Nonsignificant	→ Adhocracy (continuing)	Entropy peak	

increasing toward the expected values that would suggest an adhocratic archetype. This major breakpoint, like the one observed for week 54, was both momentous and accidental, at least in the sense that it did not constitute an explicit, deliberate effort to change Wikipedia. Nonetheless, although the large amount of content that was automatically generated by this one actor only had a temporary effect on entropy levels within the collaborative community, its significance should not be understated, as this development opened the metaphorical floodgates to new possibilities for contributions within the Wikipedia project.

In contrast, the breakpoints corresponding to weeks 142 and 204 appeared to be simple road markers. Wikipedia merely went through a slow phase of maturity, which included a relative peak in entropy.

Around week 250, Jimmy Wales admitted through the WikiEN-l mailing list that some Wikipedia content was atrocious, shaming both himself and those editors who were dedicated enough to Wikipedia to subscribe to the mailing list and see his message. Wales called upon the mailing list readers, asking what could be done to fix the problem, and their discussions spurred the most dedicated editors to take even more of the workload upon themselves to ensure that it would maintain a higher quality. This dialogue between Wales and the other editors represented a dialectic motor that drove entropy downward and, more to Wales' point, pushed the quality of articles upward. At this time, the use of "Barnstars" and the "featured article" designation, both of which had recently been introduced, dramatically increased as prevalent mechanisms to highlight high-quality editors and editorial products. Nonetheless, Wikipedia still continued to become an example of the adhocracy archetype; this revolution was just a minor adjustment along the route to that destination.

Finally, the week 335 breakpoint represented a very minor adjustment that stabilized entropy on Wikipedia and helped to nudge it toward an especially pure form of the adhocracy archetype. This breakpoint marked the point at which Wikipedia reached a stable, balanced state, so the process of structuration could be considered to be complete in its own right.

In the final chapter of this volume, we will summarize the entire story of this evolution, highlighting some of the core processes, the evolutionary patterns, and the significant interplay between leaders, adhocratic tendencies, and evolutionary processes. This concluding chapter will also reconnect the narratives of Parts I and II of this volume with one another and with the broader framework of social differentiation.

References

Boxer S (2004) Mudslinging weasels into online history. The New York Times, p 1

Carr N (2005) The amorality of Web 2.0. http://www.roughtype.com/?p=110. Accessed 29 Jan 2017

Carr N (2007) Technology: read me first: the net is being carved up into information plantations. The Guardian, p 2

Cohen N (2007) The internets, they can be cruel. The New York Times, p 4

Harkin J (2005) Saturday: lost in showbiz: big idea: The wisdom of crowds. The Guardian, p 29

Jullien N (2012) What we know about Wikipedia. A review of the literature analyzing the project(s). Available via SSRN. https://papers.ssrn.com/sol3/Delivery.cfm/SSRN_ID2308346_code728676.pdf?abstractid=2053597&mirid=1. Accessed 31 Jan 2017

Lamiell J (2014) Nomothetic and idiographic approaches. In: Teo T (ed) Encyclopedia of critical psychology. Springer, New York, pp 1248–1253

Langberg M (2005) An internet fed mostly by amateurs is frightening. San Jose Mercury News, p 1

Luthans F, Davis TRV (1981) Idiographic versus nomothetic approaches to research in organizations no. 1. Available via DTIC. http://www.dtic.mil/cgi-bin/GetTRDoc?AD=ADA104548. Accessed 31 Jan 2017

Matei SA, Dobrescu C (2011) Wikipedia's "neutral point of view": settling conflict through ambiguity. Inform Soc 27(1):40–51

Mediawiki (2016) Mediawiki history. http://www.mediawiki.org/wiki/MediaWiki_history. Accessed 29 Jan 2017

Mintzberg H (1979) The structuring of organizations: a synthesis of the research. Prentice Hall, Englewood Cliffs

Mintzberg H (1989) Mintzberg on management: inside our strategic world of organizations. Free Press, New York

Orlowski A (2005) Wikipedia founder admits to serious quality problems. The Register. http://www.theregister.co.uk/2005/10/18/wikipedia_quality_problem. Accessed 31 Jan 2017

Steiner T (2014) Bots vs. Wikipedians, anons vs. logged-ins. In: Chung C-W, Broder A, Shim K, Suel T (eds) Proceedings of the 23rd international conference on World Wide Web. ACM Press, New York, p 2014

Tkacz N (2014) Wikipedia and the politics of openness. University of Chicago Press, Chicago

Van de Ven AH, Poole MS (1995) Explaining development and change in organizations. Acad Manag Rev 20:510–540

Wales J (2001) NeutralPointOfView. http://cyber.law.harvard.edu/~reagle/wp-redux/NeutralPointOfView/982358834.html. Accessed 29 Jan 2017

Wales J (2005) [WikiEN-l] a valid criticism. http://lists.wikimedia.org/pipermail/wikien-l/2005-October/030075.html. Accessed 29 Jan 2017

Whyte E (2007) Getting to grips with acid rain. New Straits Times, p 14

Wikipedia (2005a) Wikipedia:Barnstars. https://en.wikipedia.org/w/index.php?title=Wikipedia:Barnstars&oldid=9218787. Accessed 29 Jan 2017

Wikipedia (2005b) Wikipedia:Featured article criteria: revision history. https://en.wikipedia.org/w/index.php?title=Wikipedia:Featured_article_criteria&dir=prev&action=history. Accessed 29 Jan 2017

Wikipedia (2016a) User:rambot. http://en.wikipedia.org/wiki/User:Rambot. Accessed 29 Jan 2017

Wikipedia (2016b) Wikipedia:Principles. https://en.wikipedia.org/wiki/Wikipedia:Principles. Accessed 29 Jan 2017

Wikipedia (2017a) History of Wikipedia. http://en.wikipedia.org/wiki/History_of_Wikipedia. Accessed 29 Jan 2017

Wikipedia (2017b) Wikipedia: Bots. http://en.wikipedia.org/wiki/Wikipedia:Bots. Accessed 29 Jan 2017

Wikipedia (2017c) Wikipedia: core content policies. http://en.wikipedia.org/wiki/Wikipedia:Core_content_policies. Accessed 29 Jan 2017

Wikipedia (2017d) Wikipedia: policies and guidelines. Wikipedia. https://en.wikipedia.org/wiki/Wikipedia:Policies_and_guidelines. Accessed 29 Jan 2017

Part III
Future Theoretical and Practical Directions

Chapter 11
Conclusions

11.1 Overall Narrative

The core argument of this volume can be summarized as follows. The unequal distribution of effort found in social media projects is not a mere accident, but a fact of life. A group of very productive users dominates the collaborative process due to a natural trend toward social differentiation. Contribution patterns differentiate leaders from followers relatively early in the process, which offers a convenient and flexible (adhocratic) mechanism of coordination and control through a functional response: leading means achieving by doing the most. The metaphorical "1% effect" in the title of this volume makes a direct allusion to this process. In time, online collaborative projects set aside a selected group of individuals to "make a difference." Their role is to differentiate the project from its competitors, while they, themselves, are differentiated as working leaders or, to apply a term used on several occasions in this volume, as functional leaders. Another way to understand the "1% effect" is by stating that the contribution leaders are "constitutive." Their efforts generate the momentum needed for the growth of the project.

The volume goes, however, beyond this relatively simple statement, further revealing how and by what means the process of growth takes place, including the discrete phases in the process of differentiation. For each phase, we also propose a specific type of organizational configuration exemplified by that phase: simple structure, machine bureaucracy, and adhocracy. The latter is also found to be the dominant form of organization, which is proposed as a structural framework for understanding structural differentiation in established social media.

At a more general level, this volume makes a broader point. Online collaborative organizations are, like many other human endeavors, regular affairs. They coalesce, grow, and reach an optimal state under the imperative of social structuration. Social structuration, in turn, is the product of social differentiation. Participants in online collaborative organizations assume roles that differentiate them based upon the quantity and quality of work performed, meaning that this role assumption leads to

S.A. Matei, B.C. Britt, *Structural Differentiation in Social Media*, Lecture Notes in Social Networks, DOI 10.1007/978-3-319-64425-7_11

an uneven distribution of effort. Thus, a key sign of social differentiation is the emergence of an uneven effort distribution.

Of course, an uneven effort distribution alone does not automatically translate into social organization. So, to better detect social organization, we proposed a measure that more directly taps into the meaningful significance of uneven distributions. This is social entropy, a concept that we carry over from information science. We treat social systems in a manner similar to communication systems, making the assumption that even distributions of system elements across levels of possible states indicate a lack of order (randomness) and, consequently, a maximum level of entropy. As entropy declines, some system elements occur at higher rates than others, and social order can be detected. In more tangible and directly intuitive terms, order appears when some individuals assume leadership positions by working substantially more than others and thus shaping the project to a greater degree.

This differentiation, however, does not necessitate that leading roles be accompanied by the power to order collaborators to do certain things. In other words, a leadership position does not have to have ascribed powers that explicitly differentiate "leaders" from the rest of the contributors. Working more, shaping more of the product, and intervening more in the collaboration process serve as implicit, not explicit, forms of command and control. In short, leading is doing. This is, in fact, the most common modality by which online structuration via differentiation occurs. Thus, as a corollary, we can also say that social organization online is implicit, not explicit.

Our research on the evolution of effort or collaborative entropy on Wikipedia also suggests that social order is evolutionary in nature. Group structures change over time. A claim of structural differentiation would not be complete unless it was accompanied by discrete phases in the evolution of entropy and in the uneven distribution of effort. With that in mind, when we analyzed entropy on its own, we detected discrete phases typically associated with communities of practice, including those of potential, coalescence, maturity, and stabilization.

We then enhanced this model with an organizational configuration perspective through which we proposed that Wikipedia gained its potential as a simple organization, coalesced under the guise of a quasi-bureaucracy, and ultimately matured into an adhocracy. We proposed a new method to detect the boundaries of the organizational configurations and the criteria for their identification (Chap. 9). Finally, in Chap. 10 we proposed a series of factors whose occurrence coincided with the major changes that we observed, allowing us to explain the manner in which socio-technical decisions and events outside the collaborative space impacted Wikipedia's transformation. As a whole, the introduction of policies, debates, and even interventions by automated scripts ("bots") played unique roles in shaping Wikipedia's trajectory of growth.

11.2 Elites and Social Structuration

One specific contribution of our volume was that of a dual focus that blended global and more local, elite-specific processes. Social order, after all, is a complex reality that can only be properly captured through triangulation. Consequently, social entropy or network graph measures at the global (system) level are not always sufficient. They need to be complemented by observations of the elite members' resilience over time, as social order is present to a greater degree when an uneven distribution of effort is present and some individuals are, in actuality, present at the top of the work pyramid for longer periods of time.

We measured the resilience of the top 1% contributor elite over time and learned that there is, indeed, considerable "elite stickiness" on Wikipedia. On average, across the entire period of the project, an individual identified as a top contributor in a given 5-week interval had a 40% chance of still being in the top contributor group during the following period. This stickiness, when taken over 20 weeks, was 32%, and even over 30 weeks it remained at 28%.

In a second analytic step, we used lagged regression to determine the relationship between social structuration (functional differentiation) and elite stickiness over time. The results indicated that stickiness is the product of social differentiation. As the collaborative project became more orderly and contributors were differentiated into an elite group and ordinary members, the elite also became stable. This suggests an organic process by which elites are borne out of functional differentiation, not vice versa. To put it another way, the process is bottom-up, not top-down. Elites do not precede structuration, and they do not impose it. Rather, they naturally emerge as the group becomes increasingly structured in terms of the effort contributed over time.

The function and abilities of the functional elite, however, are not extensively regulated by formal rules and procedures. Although some rules decide what content can be retained or treated as "spam" (promotional content) or vandalism (disruptive content), in the end it is the raw quantity of one's contributions that decides how much is to be preserved and who is a "leader." Further, elites do change over time, and with these changes come functional transitions between various types of organizational configurations. The most important of these is the transition to "adhocracy," a form of social organization and leadership in which roles are functional and their occupancy is temporary, acting as a function of the work that is effectively performed.

To paraphrase a dictionary definition, adhocracy does not rely on role descriptions, bylaws, and organizational charts. Work is done because it needs to be done, not because of explicit rules and norms that define who is supposed to do what and in what amount. As individual contributors come and go, and as promotion to the elite group ebbs and flows, the social structure organically reshapes itself. Thus, those who lead are those who "are there." Adhocracy, then, as a term, suggests a simple idea: adhocracy is the rule of those who are there.

The transition to adhocracy and its differentiation from several other types of organizational configurations (simple structure, machine bureaucracy, professional bureaucracy, or diversified/divisionalized organizations) was not merely assessed indirectly through our analysis of elite resilience. Rather, we captured the emergence of adhocracy by looking at a broader set of specific indicators: collaborative attractiveness and extroversion, communication flow, partner choice trends, and structural order. To measure these factors, we took a network analysis approach.

Collaborative attractiveness was represented as inbound degree centralization, while extroversion was operationalized as outbound degree centralization. We used assortativity to discern trends in partner choices among fellow collaborations, and the structure of communication flows, in turn, was modeled using betweenness centralization. Finally, structural order was captured directly through collaborative entropy.

These dimensions and measures, which map onto our set of five organizational configuration dimensions, jointly define the patterns of collaboration and contribution differentiation within the project. Assortativity, for example, captures the likelihood of elite interaction with nonelites, while the other three measures capture the degree to which individuals initiate, receive, or mediate collaborative ties. The five indicators provide a representation of the five organizational configurations articulated by Mintzberg (1979), and specific combinations at various levels of these dimensions indicate transitions from one type of organizational configuration to another. As discussed in Part II of this volume, we used this approach to determine that Wikipedia started with a simple structure, similar to a mom-and-pop business, in which work was performed in small groups of highly motivated individuals. This transitioned to a relatively short period of machine bureaucracy, characterized by a significant degree of differentiation among collaborators, and was followed by the most stable phase, that of adhocratic collaboration.

Beyond specific findings regarding the presence of or transitions between specific organizational configurations, our research converges on a proposition of greater significance: namely, online environments naturally create social structures that are organized around quasi-stable roles. The dual processes of structural differentiation and role allocation are also detectable in online collaborative environments, and through this assessment, the "mystery" of online inequality becomes easy to understand. Although online environments do invite new voices and participants into the social production game, these need not and will not end up in a state of equality. In other words, equality of opportunities to participate does not translate into equality of results. In fact, by the very logic of the measure that we used in our study, entropy, a space that is perfectly egalitarian, would also be the most random.

On a side note, our approach also suggests that the quest for equality at all costs might actually be counterproductive. All that such a blind effort will yield is maximum entropy, which is the opposite of what we should seek to create: more meaningful, functional, and structured online organizations. This issue will be developed further in future research.

More importantly, uneven distributions are not the product of mere power grabs (even though such political activities may indeed take place online), but of the simple fact that successful collaborative projects need leaders. Such leaders are not, however, authority figures in the traditional sense—that is, their leadership is not derived from externally prescribed authority bestowed through regulations, organizational charts, and strictly defined control mechanisms with minute rules that dictate who can do what and to what effect. Rather, their ad hoc authority is defined by how much they do in terms of effective work and effort spent on the project.

Likewise, advancement to an elite position is itself a function of hard work. Our results also show that working with former members of the elite is not related to one's presence in the elite at a future point in time, meaning that the elite do not reach that stature through mere cronyism. Moreover, one's authority lasts only as long as he or she continues to perform work. If leaders slack on the job, their power vanishes. There are no "idle aristocrats" and "absentee landlords" on Wikipedia or in other social media collaborative groups.

11.3 Looking Forward: Wisdom of the Differentiated Crowd

Ultimately, these findings illuminate several key points that have dramatic ramifications for modern online organizations. First and foremost, Wikipedia was never an egalitarian utopia. This is not a new finding, as a number of other researchers have come to the same conclusion (Shaw and Hill 2014), but since public misconceptions on this front existed (Lih 2009), it is worth repeating this point (Matei and Bruno 2015; Matei and Foote 2015).

As we have shown, Wikipedia is dominated at large by a small set of members who contributed the vast majority of content to the collaborative effort. Likewise, most individual articles are similarly commanded by a few key editors who largely dictate the collaboration, with many other individuals "chipping in" as necessary with comparatively minor revisions. As we and other researchers indicate, the roles are in fact very diverse and complex, which suggests that Wikipedia is a complex mechanism (Arazy et al. 2015).

Thus, the role differentiation observed on Wikipedia is a process to be reckoned with. All facets of the collaborative effort, large and small, experience stratification wherein a few individuals take charge and everyone else follows their lead. Size is largely irrelevant for this role differentiation: even the smallest elements of the community have leaders and followers, and regardless of how attractive egalitarianism may sound, the natural emergence of this social structure appears to have benefitted Wikipedia.

Despite our personal inclinations, when we work atop or within an organization, we must resist our instinct to pursue absolute equality among all contributors. Role differentiation, unromantic though the notion may be, is important for organizational success, so leaders should foster the emergence of such distinctions among members rather than seeking homogenization. This is just as true in the online realm as it is offline, even if other aspects of collaboration may change.

11.4 The Fallacy of the Mini-Wikipedia

As is often forgotten, large-scale collaborative groups, as with organizations of any nature, did not merely emerge in years past as smaller-scale versions of themselves that are otherwise identical to their current forms. Case in point: Wikipedia, in its earliest days, was not the adhocracy that we observe now. It instead functioned akin to a small business, like the classic mom-and-pop store, and only when it outgrew those constraints could it transcend its initial configuration—though, unlike most major corporations, Wikipedia eschewed some tendencies toward a bureaucratic organization and instead found stability as the adhocracy that we know today.

This is not a trivial realization. We are frequently driven to compare current individuals, groups, and products against their predecessors, assuming that the best of tomorrow will look like a smaller, younger clone of the gold standard from today. Is LeBron James the next Michael Jordan? What company will be the next Walmart, or the next Chipotle? Who is the next *American Idol* (following in the footsteps of Kelly Clarkson and Carrie Underwood, the "gold standards" of popular and financial achievement, rather than the other not-as-successful winners)?

But in looking for the "next Wikipedia," we should not attempt to find collaborative communities that already resemble the Wikipedia of today. Those organizations most likely to find similar prosperity as Wikipedia are those that presently operate as budding small groups, with a growing membership that has the potential to transcend their simple structures and grow along the same path as their predecessor. In other words, organizations aiming to emulate Wikipedia must follow in its footsteps rather than attempting to start at the end point of Wikipedia's journey. It is not enough to merely make a "mini-Wikipedia" based only on what we observe today and then to just sit back and wait for contributors to join and fill the void.

11.5   Embracing Revolution

Just as it is important to trace the footsteps of Wikipedia's growth, there are lessons to be drawn from the jumps between those footsteps, as well. As our research has shown, Wikipedia proceeded through several distinct phases in its first several years, each of which was marked by a moment of change that served to redefine the collaborative community as a whole. In other words, Wikipedia's growth was not solely one of gradual, incremental evolution. It also involved revolutionary shifts that fundamentally and permanently redefined the organization as a whole.

This is a key point for leaders of online startup organizations to recognize as they chart the growth of their respective endeavors over time. As Wikipedia's leaders realized, for instance, while the system architecture that initially facilitated contributions was sufficient in its earliest days—when, from an organizational standpoint, it still functioned as a simple structure—technical limitations eventually became a force that inhibited further growth. Only once the existing editing and publishing engine was replaced was Wikipedia able to grow beyond its initial conception as an experiment and support a long-term community of members (Britt 2013; Mediawiki 2016).

Not coincidentally, this was also the point at which Wikipedia's configuration transitioned from its original path, which it had been following toward a machine bureaucracy, and began moving toward an adhocratic system instead. In other words, social and organizational change online may be accompanied by technological transformation. Yet technology is not destiny, as social and organizational forces are more important, to a certain extent, than technological ones—after all, if Wikipedia's users were bound by their technologies with no agency of their own, then the site would still be using its Phase I engine, and we likely would not be talking about the collaborative group today.

The most important takeaway from this is that organizational leaders need to be ready and willing to catalyze (or at least enable) dramatic, earthshaking changes in their organizations when they are ready to grow. As an organization expands, it will inevitably encounter barriers that inhibit further growth. A wise leader will be ready to break down those barriers, freeing the organization to continue to grow.

Furthermore, such growth will not always be expected, and in some eyes, it may not even be desirable. For example, the abandonment of the first technological infrastructure, the so-called UseModWiki editing and publishing engine (Wikipedia 2016), and the adoption of the increasingly complex MediaWiki platform (Wikipedia 2017), while a necessary step in Wikipedia's development, also facilitated later developments like the introduction of large-scale editing bots. These bots, computer-automated editing tasks triggered by the addition, change, or presence of specific types of article content or editing behaviors, made Wikipedia an even more complex socio-technical laboratory. At the time of the first bot's activation in 2002, however, many members viewed the use of automated editing software like "rambot" as a potential source of devastation that could drive the growing project to ruin. Bots quickly became a hot-button issue in the community, particularly due to the significant technical issues that sometimes resulted from their use. Ultimately, discussions among concerned members resulted in the creation of a formal bot policy that allowed the continued use of bots but set limits on their application in order to prevent Wikipedia from descending into chaos.

Today, bots play a vital role on Wikipedia, performing a wide range of tasks such as cleaning up articles so that they meet Wikipedia's presentation standards and mining external databases in order to create new articles and populate existing ones with content. Yet permitting bots, while an important step for Wikipedia's development, was one that filled many Wikipedians with trepidation at the time. Still, regardless of one's fears that an idyllic organization may be ruined by dramatic changes, it is often necessary to embrace those changes and foster revolution in order to allow the organization to reach its potential. The old saying, "If it ain't broke, don't fix it," no longer applies.

This, however, does not mean that revolution should be incited for its own sake. As we have shown, Wikipedia largely stabilized midway through its seventh year and has not undergone any substantial revolution since then; for that matter, even those revolutionary moments that we detected after 2003 were of much less consequence than those that came before. Furthermore, even bot contributions are a relatively minor part of the English Wikipedia's editorial life today. Not factoring in edit size, only 5% of individual edits are performed by bots (Steiner 2014), and given that many such automated edits constitute minor stylistic adjustments, their overall workload is quite small compared with the community of human editors. Therefore, it is not as though the introduction of bots served to reinvent the entire project; it only altered its potential and its course. So while the revolution was certainly a meaningful change, it should not be viewed as the abandonment of everything that had made Wikipedia successful and interesting in the first place.

All told, Wikipedia has reached a state of relative stability, and it remains one of the most-visited websites in the world. While it is important to avoid stagnation, it is also necessary to consider what benefits a revolutionary change could provide. Moreover, adhocratic regimes are notoriously hard to change, as they tend to function by customs and practical know-how. With that in mind, for many years, there has not been compelling evidence to suggest that major alterations to Wikipedia would provide any real benefits, and until there is such a rationale, there is no reason for Wikipedia's leaders to attempt to affect the existing system in such a dramatic fashion. In short, change is good, but only if it is purposeful change.

11.6   Interchangeable, Irreplaceable Leaders

Perhaps the most curious finding to date is the stickiness of Wikipedia elites. Those who contribute the most content to Wikipedia in a given week are very likely to continue to find themselves in the upper echelons in the following week, and the week after that, and so forth, to a far greater degree than what we would expect by random chance alone. As previously mentioned, when we consider contributions made during successive 5-week periods, elite stickiness from one interval to the next was 40%.

This is especially curious when one considers that the roster of elite members today is nonetheless very different than its constitution was several years ago. In other words, top-level contributors remain in that role for a time, but gradually recede into the shadows, allowing others to take the helm in their absence. Nonetheless, as we have seen, there is a consistently stark contrast between the massive amount of content contributed by those elites and the comparatively middling contributions of all other editors. This holds true regardless of whether we compare the elites against the masses throughout Wikipedia's lifetime, as we did in Chap. 5, or whether we isolate a specific period of time (such as all contributions made only in a particular week).

All of this ultimately suggests that, for Wikipedia and other collaborative systems of a similar nature, it is essential to have a subset of contributors who perform the bulk of the work, carrying the effort forward largely on their own shoulders. Such members are those who have sufficient experience to understand the way in which Wikipedia operates and the "best practices," in a sense, of contributing. As such, these individuals are able to lead by example, passing down their knowledge to up-and-coming editors, and eventually passing the metaphorical baton of leadership—but not before leading long enough to first confer their own wisdom, both in terms of the organizational behaviors that they model and the raw factual knowledge that they add to Wikipedia.

In short, Wikipedia needs to have leaders in order to survive, and those leaders have to do the lion's share of the work—as previously noted, by our calculations, the top 1% of contributors provided 77% of Wikipedia's content. Those leadership positions are indeed sticky, but they are not permanent. As such, those leaders can and do gradually rotate, with new leaders eventually stepping in to fill the role. In other words, the role of "elite" is a crucial one for Wikipedia, which illustrates the need for differentiation between editors at large and those who adopt positions at the top of the system, even if the particular individuals who fill those roles vary over time.

Furthermore, if the constitution of this elite group changed too rapidly, dramatically shifting from one day to the next, then the system would likely be thrown into chaos, with little in the way of guiding principles to consistently direct behavior. But because elites tend to remain in the top levels of Wikipedia for at least a few weeks, their resilience in that position lends a degree of stability to the system that they help to drive. Even as leaders gradually cycle in and out of the role, the transition periods are gradual enough to foster an air of consistency among Wikipedia contributors. As such, while their spot among the elites is only quasi-permanent, those who lead for a short time are nonetheless irreplaceable—at least until their stickiness reaches its end.

11.7 Structures and Structuration Through an Entropy Lens: What Is the Optimal Range?

Although this volume presents two different approaches to detect structural differentiation and changes in its composition, the first one, which relies more heavily on entropy, is simpler and may gain more currency as a result.

With that said, measuring structuration through entropy values presents both advantages and disadvantages. The main advantage is parsimony. A single statistical indicator provides a very specific measurement through which an organization's state may be pinpointed. The main disadvantage, then, is the linear progression embedded in the measurement. While it is valid to assume that maximum entropy values represent a state of disorganization, is it appropriate to claim that the minimum level of entropy is the most perfect form of an organization? Recall that entropy exhibits a level of zero only when a single element accounts for all of the activity within the system. Considering the collaborative work on Wikipedia, if one member singlehandedly accounted for all the contributions, would that make Wikipedia the most efficiently organized system in the world, online or offline? Insofar as social systems are collaborative, of course not. To be blunt, the example is nonsensical at its core, since no collaborative system of any significant size can offload all of the work onto one individual.

Let us therefore move a bit closer to reality. If a very small group of individuals—let us say, 10 or 50 members—accounted for all the work performed in a collaborative project of 1,000,000 members, would that reflect a more "organized" and socially differentiated collaborative space than one in which 100 or 1,000 members contributed the same amount of work? Intuitively, one might be tempted to say that a Wikipedia edited by ten individuals would still be closer to an ideal state of differentiation than one edited by 100. However, it is, of course, somewhat more realistic than a Wikipedia edited by one individual.

As we move down the ladder of unevenness, we realize that although grotesquely small values are not to be taken seriously, we still have a hard time naming the magic threshold at which a relatively low (yet still reasonable) entropy value indicates that we have entered the realm of appropriate "social order." Does such a figure exist? Perhaps not, or at least there is no such number that we can uncover through deduction alone. However, it is certainly possible to use inductive research on a range of values to ascertain the entropy levels that designate various levels of social organization. If an organization exceeded the maximum value of this entropy range, its social order would unravel, while if entropy dropped under the minimum, social order would ossify into a system dominated by very few, which nullifies the entire idea of social order based on collaboration. In other words, empirical research might reveal that entropy, as a measure, captures social order in a curvilinear manner. At the lower and higher ends, social order is a vanishing reality. Social "noise" might drown it at one end of the spectrum, while rigid control by very few would smother it at the other end.

In view of this, entropy-based research on social order needs to go through an extensive process of validation and calibration. In a way, using entropy as a social diagnosis instrument is similar to using a thermometer in medicine. While we know from our own experience that the temperature of the human body cannot be too low or too high, the thresholds for what is "low" and what is "high" could not have been and were not determined by deductive insight alone. Rather, a significant amount of empirical research of a large number of humans, in a variety of conditions, leads to the conclusion that the healthy human body temperature cannot vary more than fractions around 37 °C (98.6 °F) degrees (Bleich et al. 1978).

The calibration of the normal temperature range of the human body was a major medical accomplishment, providing a significant tool for detecting and tracking febrile states and, through them, the evolution of certain illnesses. With that said, the observed values do vary, even if ever so slightly. For example, the human body is a bit colder in the morning, a change that is not associated with illness but with the natural fluctuations inherent in an otherwise healthy stable state. Furthermore, measuring and charting body temperature did not immediately provide a direct answer about the root cause of a given illness. It took many decades until fever was associated with various conditions, such as epidemics triggered by infectious agents, and it took even longer for researchers to provide a specific physiological explanation for the infection-related mechanisms that generate the directly observable symptom of fever.

At present, entropy-based research on social order is at a crossroads similar to that when physicians first began to realize that normal body temperatures vary among specific values and that significant departures suggest a pathologic problem. The next step, then, is to determine the reasonable and expected ranges for collaborative entropy in a variety of human organizations. Importantly, the variability of these ranges might be far greater than those associated with the medical analogy presented above. After all, organizations involving innumerable human actors may be more challenging to predict than a single person, and organizations in distinct contexts or with varying internal features may be quite different from one another, as opposed to the relatively standard biology of human beings. It is even possible that these ranges might not even overlap across contexts. Furthermore, since entropy is dependent on determining the elements of the system and the manner in which they manifest themselves in relation to a given variable of interest, it might not even be possible to compare entropy values and ranges in some situations. Beyond this, even when "healthy" ranges are eventually determined for each context, much more research will be necessary for us to outline the system-specific "physiological" mechanisms that are responsible for entropy variations.

Therefore, entropy measurement can and should be improved at the observational level as well. As our research currently stands, we only take into account the amount of work completed to determine participant roles, and those are defined dichotomously as leader vs. regular contributor. However, more recent work has provided ways through which we may detect more nuanced patterns of interaction and role allocation that make qualitative distinctions between substantive and formal edits and between broad and deep contributions (Arazy et al. 2015). We can easily imagine a modified entropy indicator that takes into account both the nature of the role and the breadth or depth of the work to provide a richer, multidimensional measurement of social order, one that implicitly controls for the specifics of the work performed. Of course, this would cause us to lose some of the parsimoniousness of the unidimensional metric, but it may nonetheless be a worthwhile sacrifice for the sake of improving our understanding and classification of contributor roles.

Likewise, our entropy-based methodology may open other new avenues for research as well, as it provides a common theoretical and methodological ground through which we may unify and explain a wide variety of problems and processes related to social order. One particularly important and promising area of investigation would be the effect of optimal ranges of entropy on system states and system outcomes. The question of whether optimal entropy ranges translate into optimal system functioning is particularly tantalizing. Are systems that hold an optimal or "normal" entropy level more efficient? Or, in more intuitive terms, at what level of unevenness and functional distribution of roles do systems allocate tasks and allow individuals to function most efficiently? Within what entropy ranges will the quality of the collaborative product be maximized?

If, let us say, Wikipedia reaches a certain entropy optimum that, as determined by repeated observations, is found to be most efficient, will the content reach its highest level of quality as a result? Will the articles become especially complete, timely, accurate, or well written? Will that optimized entropy level instead drive the quantity of contributions rather than its quality? Or will it ultimately yield some balance between the competing desires of quality and quantity? Again, this type of deep research has yet to be conducted. Some of these issues are already undergoing early investigations (e.g., Sydow et al. 2016), and we sincerely hope that further research in this vein will provide some valuable insights in this direction. Our own upcoming research, which includes Wikipedia projects in non-English languages (French, Spanish, Romanian, Arabic, etc.), will more directly examine these issues and will both broaden and expand upon the foundational work that was presented in this volume.

References

Arazy O, Ortega F, Nov O, Yeo L, Balila A (2015) Functional roles and career paths in Wikipedia. In: Cosley D, Forte A, Ciolfi L, McDonald D (eds) Proceedings of the 18th ACM conference on computer supported cooperative work & social computing. ACM Press, New York

Bleich HL, Boro ES, Dinarello CA, Wolff SM (1978) Pathogenesis of fever in man. New Engl J Med 298(11):607–612

Britt BC (2013) Evolution and revolution of organizational configurations on Wikipedia: a longitudinal network analysis. Dissertation, Purdue University

Lih A (2009) The Wikipedia revolution: how a bunch of nobodies created the world's greatest encyclopedia. Hyperion, New York

Matei SA, Bruno RJ (2015) Pareto's 80/20 law and social differentiation: a social entropy perspective. Public Relat Rev 41(2):178–186

Matei SA, Foote J (2015) Transparency, control, and content generation on Wikipedia: editorial strategies and technical affordances. In: Matei SA, Russell MG, Bertino E (eds) Transparency in social media: tools, methods and algorithms for mediating online interactions. Springer, New York, pp 239–253

Mediawiki (2016) Mediawiki history   http://www.mediawiki.org/wiki/MediaWiki_history. Accessed 29 Jan 2017

Mintzberg H (1979) The structuring of organizations: a synthesis of the research. Prentice Hall, Englewood Cliffs

Shaw A, Hill BM (2014) Laboratories of oligarchy? How the iron law extends to peer production. J Commun 64(2):215–238

Steiner T (2014) Bots vs. Wikipedians, anons vs. logged-ins. In: Chung C-W, Broder A, Shim K, Suel T (eds) Proceedings of the 23rd international conference on World Wide Web. ACM Press, New York, p 2014

Sydow M, Marcin S, Katarzyna B, Paweł T (2016) Diversity of editors and teams versus quality of cooperative work: experiments on Wikipedia. J Intell Inf Syst. doi:10.1007/s10844-016-0428-1

Wikipedia (2016) UseModWiki. https://en.wikipedia.org/wiki/UseModWiki. Accessed 29 Jan 2017

Wikipedia (2017) MediaWiki. https://en.wikipedia.org/wiki/MediaWiki. Accessed 29 Jan 2017

Appendix A: Data Collection, Management, Preprocessing, and Analysis

Data Acquisition

In order to examine the research questions highlighted in this volume, we examined every revision made to every encyclopedic article on Wikipedia between January 16, 2001 and July 5, 2010. Luca de Alfaro and his colleagues at the University of California, Santa Cruz contributed the complete data set of revisions to the Wikipedia corpus made during this period. As part of their WikiTrust project conducted in conjunction with the Wikimedia Foundation, de Alfaro, Adler, and Pye (2010) parsed all revisions to all Wikipedia articles and collected a wide range of data about each of those revisions. The parsing was done only once, due to computing space limitations, in 2010.

During the 9.5-year interval for our study, a total of 22,792,885 users contributed 235,701,162 million revisions to 7,296,360 different articles. Of these users, 3,112,211 were logged into a registered account when making their revisions and could therefore be tracked as individuals, while the remaining 19,680,674 users made changes anonymously. Anonymous users were only identified by the IP address of the computer that was used to perform the edit. Thus, anonymous users cannot be identified as unique individuals—if a Wikipedia editor's computer happened to be assigned an IP address previously used by a fellow editor, their contributions could overlap and be incorrectly identified as coming from the same individual using the same IP address. Likewise, a single anonymous user could be assigned any number of IP addresses over time, and furthermore, a given individual with a registered account could have made some revisions while logged in and contributed others anonymously, thus creating the illusion of an additional anonymous user.

With that in mind, although anonymous "users" were six times more numerous than logged-in users, they only accounted for about one-third of the total amount of contributions, as measured by de Alfaro et al.'s delta formula (see below). This makes sense, as it would be difficult for others to form long-term relationships with a fellow editor who remains anonymous and therefore does not have an easily

© Springer International Publishing AG 2017   179
S.A. Matei, B.C. Britt, *Structural Differentiation in Social Media*, Lecture Notes
in Social Networks, DOI 10.1007/978-3-319-64425-7

identifiable (or even a unique or stable) username. In contrast, when an individual registers an account on Wikipedia and logs into that account before editing a Wikipedia article, that individual is choosing to be part of the Wikipedia editing community and to be recognized based on the contributions that he or she has made, all of which can be connected back to the username.

Thus, for the purposes of this project, we focused on the evolution of contribution diversity and entropy only among the 3,112,211 logged-in users, who jointly offered about two-thirds of the total effort in the production of Wikipedia content between January 16, 2001 and July 5, 2010. This allowed us to minimize the chance of misrepresenting one editor as multiple users or vice versa and to properly assess the activities of the most dedicated contributors to the project: those who bothered to sign up.

Beyond this, the data was bounded in time due to the nature of the analysis, which specifically looked at effort. Our collection of Wikipedia revisions needed to be expressed in a manner that demonstrated the system of collaborative relationships as it changed over time. Therefore, the raw data was preprocessed by de Alfaro and his colleagues and expressed as a new unit of measurement signifying effort.

More specifically, the de Alfaro parser was used to analyze each editorial intervention and calculate the amount of effective change resulting from a given revision and, by proxy, of effort invested in the change. Due to the specific method of calculation, the value is also called "delta," which makes allusion to the fact that effort is a human facet of the concept of "edit distance."

de Alfaro's formula (Eq. A.1) takes into account all three kinds of changes that could be recorded for a given intervention: additions, deletions, and changes to existing text. In so doing, the formula limits the impact of deletions and mere copy-paste changes, thereby offering a superior measurement strategy over many previous attempts in similar studies that typically used mere character counts to assess the significance of a given contribution.

$$d(u,v) = \max(I,D) - 0.5\min(I,D) + M, \text{where } I = \text{Total inserted text,}$$
$$D = \text{Total deleted text, } M = \text{Total relative change of position}$$
$$(\text{a measure of text reordering}).$$

$$(A.1)$$

The d score was used throughout this study as the main measure of user contribution (effort). Delta (effort) scores were used in all calculations of core measures and relationships, including entropy, the presence of a given individual in the contributing elite, and coeditorial connections between participants in the collaborative effort. Inequality was measured using a standard application of the entropy formula, as detailed in Chap. 5.

We chose entropy over other measures of inequality, such as the Gini coefficient, due to its broader conceptual implications. Lower levels of entropy can, in fact, be interpreted as a higher level of system structuration. In other words, as entropy goes down, we expect that the group will become more structured.

Network Conception

In Part II of this volume, organizational configurations were assessed using five different dimensions, four of which were network dimensions. In order to serve this objective, we undertook a second preprocessing phase during which the effort data was organized into a series of networks to describe the evolution of collaborative connections and self-organizing practices among Wikipedia editors over time.

In order to assess these changes, Wikipedia's life span was first divided into seven-day windows, beginning from the first week after its formal launch on January 15, 2001. These windows were narrow enough to yield many data points over the longitudinal data set but wide enough so that the network constructed for each week would show a rich, meaningful network of interpersonal connections that developed over those 7 days, given the number of revisions made during that period.

The first network, therefore, described all collaborative connections formed during Wikipedia's first week of activity, the second network reflected those connections made during second week, and so forth. The sequence of these networks, then, described changes in the community's collaborative processes over time. These networks were conceptualized in terms of user-to-user (editor-to-editor) collaborative connections, or the collaborative ties formed when one user changed content that another user had contributed. In other words, when two different individuals engaged with the same material, a coeditorial connection between the two users was generated (Matei et al. 2015).

However, precise data on the exact text that each user contributed and changed could not feasibly be analyzed across the entire Wikipedia corpus for its development over 9.5 years. In other words, while it was feasible to analyze data regarding which users revised which articles at what points in time, more detailed assessments of user changes to the exact sections and words within an article were not computationally feasible.

Consequently, even if two individuals revised the same article, it was possible that they altered entirely distinct portions of the text and never actually interacted with one another's work. Likewise, it was also possible that one editor could revise a given article and then have that change reverted or otherwise heavily modified by subsequent editors. In that instance, someone viewing and editing the article at a later date might never even see the contribution of the first editor, as the text that person contributed might have already been erased or modified beyond recognition.

Given that coediting activity could not be directly assessed, a modified version of Isard's (1954) gravity model of trade was used as a proxy measurement to instead evaluate the likelihood that any two editors interacted (Matei et al. 2015). Isard's original model was designed to assess the likelihood of two countries, provinces, or other distinct entities sharing a trade relationship based on each entity's trade activity and the geographic distance between them. For any two communities, their shared score under Isard's formulation increased as each group's trade activity increased, but declined as they grew further apart. The higher the result based on

Isard's formula, the more likely that the two groups shared a trade relationship, and the more likely that relationship was to be especially active as well.

Based on this formulation, as well as prior work on Wikipedia (Britt 2011, 2013; Matei et al. 2015), the present study used the gravity model of online interaction (see Eq. A.2 below) to assess the likelihood and extent of coediting interactions between any two Wikipedia editors. In this case, the number of editing iterations between two contributors' revisions represented the "distance" between two revisions to the same article. The more revisions made between the contributions of those two editors, the more likely it was that the first editor's work had been significantly altered or completely removed before the second editor ever approached the article.

$$F_{ij} = \frac{M_i M_j}{D_{ij}^2}, \text{where } M_i \text{ represents the quantity of change made}$$

by the first revision, M_j represents the change from the second

revision, and D_{ij} represents the difference between revision numbers

i and j for the given Wikipedia article

(A.2)

The significance of changes made to a given article also played a major role in the likelihood that two editors interacted with the same text (Britt 2011; Matei et al. 2015). A user who added several pages of text was much more likely to have engaged with the work of many previous collaborators—and to have many subsequent editors engage with his or her own work—than another user who added only a few words or a single sentence. As mentioned above, the significance of any given revision was quantified by de Alfaro et al.'s (2010) "delta" measure, which was based upon the number of words added, removed, changed, and moved across a given article.

In tandem, the significance of any two revisions and the number of iterations between them gives us likelihood that the two editors themselves interacted with the same text in the article. Thus, the weight of a network edge formed between two editors who revised the same article represented the probability of their interaction, so the combination of all of these edges yielded a probabilistic network of collaboration, a concept that was proposed through the Visible Effort project (Matei et al. 2015).

These relationships formed the basis of the weekly networks designed to demonstrate changes in the collaborative system on Wikipedia over time. For each revision j made to any article on Wikipedia, a directional edge was formed from the author of that revision to the authors of all previous revisions of the same article. This edge was placed within the weekly network whose seven-day time range contained the point at which the revision j was made, as determined by its timestamp. The weight, or degree, of the edge pointing from the author of revision j to the author of any prior revision i within the same article was calculated using the gravity model of online interaction.

In the event that the weekly network in question already contained an edge pointing from the author of revision j to the author of revision i, then the weight of the new edge to be created was simply added to the weight of the existing edge. This produced a single edge describing the engagement that the author of revision j devoted within the week in question to the work of the editor who created revision i. As such, a coediting relationship between two users could be strengthened through repeated revisions to the same article or through mutual collaboration on multiple Wikipedia articles.

It is worth making special note that any network edge formed between any two editors did not carry over to subsequent networks describing later weeks of interaction on Wikipedia. This stands in contrast with the networks devised in Britt's (2011) previous study on Wikipedia's growth over time. In that earlier study, networks aggregated across time, so any given week's network consisted of all the collaborative connections formed in that week in addition to those formed in all prior weeks. As such, each weekly network was larger than the one that preceded it, highlighting the growth of collaborative connections over time.

In the present study, all edges formed in prior weeks were instead ignored in constructing the network for any subsequent week. Because edges were not aggregated across weekly networks, organizational structures that may have been prominent months or years earlier did not necessarily persist—such persistence of long-forgotten structures would have needlessly obscured later structural changes. In other words, keeping the weekly networks distinct permitted more direct observation of sharp changes in the way that the community behaved from one week to the next.

In total, 491 networks were generated to represent the collaborative networks (see Table A.1). The first 3 weeks as well as the final week in the data set were not included in the network analyses due to issues with data size and completeness (Matei et al. 2015; see also Table A.2).

The entropy of contributions, which was the fifth dimension used to assess organizational configurations, was also assessed at a weekly level for a total of 495 weeks—all of which were used for the analyses in Part I of this volume (see Table A.3) and 491 of which were included in the analyses in Part II (Table A.2). This entropy value was normalized such that the range of possible values spanned the interval between 0 and 1 in order to ensure comparability over time. The entropy of each week took into account all the contributions made by all Wikipedia editors up to that point. Thus, this represented a cumulative entropy measure, so the final entropy value represented the state of entropy (and diversity) for the entire Wikipedia project over the entirety of the study interval.

Data Management

Because of the massive size of the data set in this study—the raw data comprised a text file of over 67 GB—sophisticated hardware and software resources had to be utilized to handle the data. The raw data was stored on and accessed from the Coates cluster within Purdue University's Community Cluster Program, a supercomputing

Table A.1 Weekly descriptive statistics for Wikipedia activity

Weekly number of revisions	Cumulative number of articles revised	Weekly number of articles revised	Authors	Total cumulative delta (effort)	Week beginning date
45	40	40	3	740	1/16/01
123	133	97	6	11,409	1/23/01
165	261	132	6	16,425	1/30/01
109	345	93	7	20,289	2/6/01
132	433	98	9	23,193	2/13/01
221	571	165	12	29,035	2/20/01
110	648	88	12	44,462	2/27/01
335	870	250	17	73,043	3/6/01
160	982	120	22	80,660	3/13/01
284	1,152	233	26	114,102	3/20/01
216	1,308	175	32	161,180	3/27/01
136	1,412	110	32	175,537	4/3/01
137	1,519	116	35	196,585	4/10/01
254	1,742	229	36	203,656	4/17/01
201	1,924	195	38	219,766	4/24/01
288	2,178	273	40	223,173	5/1/01
249	2,379	218	46	241,421	5/8/01
507	2,781	437	47	288,224	5/15/01
365	3,072	321	47	329,671	5/22/01
227	3,249	200	50	364,299	5/29/01
152	3,357	130	50	372,386	6/5/01
186	3,475	142	52	395,248	6/12/01
151	3,560	101	54	399,290	6/19/01
252	3,700	192	54	427,332	6/26/01
169	3,813	145	56	443,159	7/3/01
134	3,915	120	57	457,911	7/10/01
159	4,035	137	63	470,816	7/17/01
795	4,557	593	97	530,002	7/24/01
527	4,940	449	109	589,631	7/31/01
475	5,267	383	116	666,843	8/7/01
537	5,634	442	125	755,122	8/14/01
579	6,075	484	130	811,860	8/21/01
496	6,428	414	131	887,242	8/28/01
638	6,850	496	140	964,142	9/4/01
875	7,421	647	145	1,086,221	9/11/01
961	8,026	723	158	1,219,634	9/18/01
1,847	9,264	1,470	204	1,526,448	9/25/01
1,174	9,982	879	221	1,729,675	10/2/01
1,641	11,068	1,321	229	2,130,349	10/9/01
1,412	11,877	1,005	235	2,402,473	10/16/01

(continued)

Table A.1 (continued)

Weekly number of revisions	Cumulative number of articles revised	Weekly number of articles revised	Authors	Total cumulative delta (effort)	Week beginning date
1,512	12,780	1,104	257	2,647,717	10/23/01
1,897	13,793	1,270	278	2,993,555	10/30/01
1,573	14,623	1,053	291	3,320,229	11/6/01
2,263	15,272	1,057	316	3,820,575	11/13/01
3,353	16,070	1,415	331	4,140,711	11/20/01
3,846	16,738	1,479	357	4,394,154	11/27/01
4,464	17,523	1,689	388	4,858,662	12/4/01
4,341	18,158	1,514	421	5,437,767	12/11/01
2,900	18,598	1,115	431	5,676,836	12/18/01
3,071	18,975	1,122	445	5,786,348	12/25/01
3,096	19,312	1,078	460	5,881,303	1/1/02
3,819	19,856	1,384	474	5,997,379	1/8/02
3,134	20,273	1,101	493	6,087,116	1/15/02
2,695	20,684	1,198	514	6,200,101	1/22/02
2,755	21,138	1,511	536	6,277,567	1/29/02
3,396	21,740	1,796	558	6,372,883	2/5/02
3,362	22,204	1,666	575	6,512,972	2/12/02
35,821	32,240	32,046	646	6,981,488	2/19/02
4,774	33,252	2,559	667	7,119,028	2/26/02
4,407	34,036	2,157	685	7,300,457	3/5/02
4,396	34,771	2,119	705	7,530,141	3/12/02
4,609	35,509	2,043	728	7,737,530	3/19/02
4,223	36,296	2,048	748	7,938,906	3/26/02
5,036	37,109	2,694	770	8,162,448	4/2/02
4,287	37,801	2,446	782	8,581,353	4/9/02
3,303	38,334	1,908	795	8,725,646	4/16/02
2,737	38,837	1,607	807	8,900,768	4/23/02
2,845	39,383	1,684	820	8,997,628	4/30/02
2,242	39,720	1,214	832	9,072,738	5/7/02
4,014	40,287	2,689	843	9,193,510	5/14/02
5,593	41,144	3,146	858	9,381,343	5/21/02
4,860	41,987	2,703	874	9,566,524	5/28/02
5,587	42,923	2,841	896	9,795,261	6/4/02
7,414	44,549	4,227	918	$1.01E^7$	6/11/02
6,542	45,659	3,446	935	$1.05E^7$	6/18/02
3,972	46,280	2,305	949	$1.06E^7$	6/25/02
5,019	47,125	2,981	972	$1.08E^7$	7/2/02
5,031	48,101	2,629	987	$1.10E^7$	7/9/02
6,034	48,921	3,369	1,007	$1.13E^7$	7/16/02
9,573	50,445	4,876	1,044	$1.16E^7$	7/23/02

(continued)

Table A.1 (continued)

Weekly number of revisions	Cumulative number of articles revised	Weekly number of articles revised	Authors	Total cumulative delta (effort)	Week beginning date
9,680	52,028	4,670	1,098	$1.20E^7$	7/30/02
11,429	53,840	5,496	1,124	$1.24E^7$	8/6/02
9,859	55,266	4,937	1,149	$1.27E^7$	8/13/02
11,995	57,008	6,495	1,186	$1.31E^7$	8/20/02
14,895	59,342	8,368	1,224	$1.39E^7$	8/27/02
14,931	61,474	7,521	1,271	$1.45E^7$	9/3/02
15,884	64,117	8,593	1,307	$1.52E^7$	9/10/02
16,942	66,511	8,791	1,389	$1.60E^7$	9/17/02
20,091	72,158	11,686	1,444	$1.77E^7$	9/24/02
14,847	74,636	7,856	1,508	$1.82E^7$	10/1/02
15,294	76,555	8,607	1,562	$1.86E^7$	10/8/02
31,372	94,590	23,715	1,616	$2.56E^7$	10/15/02
29,808	112,010	22,613	1,658	$3.23E^7$	10/22/02
12,118	113,459	5,861	1,719	$3.27E^7$	10/29/02
12,242	114,831	6,408	1,766	$3.32E^7$	11/5/02
12,953	116,591	6,224	1,824	$3.36E^7$	11/12/02
11,557	118,824	6,073	1,878	$3.43E^7$	11/19/02
10,709	120,283	5,394	1,916	$3.46E^7$	11/26/02
16,013	121,786	8,011	1,974	$3.51E^7$	12/3/02
32,036	123,381	25,994	2,023	$3.62E^7$	12/10/02
23,083	124,922	15,868	2,079	$3.67E^7$	12/17/02
11,611	126,202	5,173	2,132	$3.72E^7$	12/24/02
17,181	128,071	7,513	2,209	$3.79E^7$	12/31/02
15,696	129,727	7,059	2,250	$3.85E^7$	1/7/03
16,657	131,703	7,195	2,327	$3.91E^7$	1/14/03
18,248	133,973	8,444	2,492	$3.96E^7$	1/21/03
14,532	136,290	7,119	2,643	$4.01E^7$	1/28/03
14,392	137,821	6,742	2,756	$4.07E^7$	2/4/03
16,659	139,655	7,614	2,842	$4.13E^7$	2/11/03
18,689	141,720	8,827	2,925	$4.20E^7$	2/18/03
15,977	144,022	7,803	2,992	$4.26E^7$	2/25/03
15,865	145,908	7,404	3,063	$4.32E^7$	3/4/03
14,915	147,848	6,950	3,138	$4.39E^7$	3/11/03
15,992	149,670	7,188	3,197	$4.44E^7$	3/18/03
17,625	151,605	7,961	3,255	$4.49E^7$	3/25/03
15,050	153,437	6,838	3,300	$4.53E^7$	4/1/03
18,589	155,560	8,272	3,372	$4.60E^7$	4/8/03
16,571	157,637	7,816	3,448	$4.65E^7$	4/15/03
18,816	160,062	9,015	3,512	$4.72E^7$	4/22/03
19,072	162,349	9,012	3,579	$4.78E^7$	4/29/03

(continued)

Table A.1 (continued)

Weekly number of revisions	Cumulative number of articles revised	Weekly number of articles revised	Authors	Total cumulative delta (effort)	Week beginning date
17,908	164,542	8,709	3,668	4.86E^7	5/6/03
22,104	166,878	10,572	3,788	4.93E^7	5/13/03
22,195	169,510	10,578	3,911	5.01E^7	5/20/03
21,290	172,156	10,352	4,063	5.08E^7	5/27/03
22,198	174,473	10,737	4,181	5.15E^7	6/3/03
22,601	176,912	10,644	4,318	5.22E^7	6/10/03
26,029	179,321	15,527	4,450	5.30E^7	6/17/03
19,580	181,569	10,212	4,543	5.36E^7	6/24/03
25,393	187,145	13,543	4,676	5.62E^7	7/1/03
22,002	189,716	11,069	4,856	5.70E^7	7/8/03
23,280	192,486	11,458	5,069	5.78E^7	7/15/03
27,989	194,961	15,332	5,277	5.87E^7	7/22/03
25,118	197,418	12,232	5,474	5.94E^7	7/29/03
29,848	201,254	13,438	5,722	6.04E^7	8/5/03
28,912	204,274	13,169	5,906	6.14E^7	8/12/03
28,758	207,363	13,655	6,059	6.23E^7	8/19/03
28,168	210,306	13,154	6,214	6.33E^7	8/26/03
28,003	213,257	13,287	6,408	6.41E^7	9/2/03
22,568	215,980	10,788	6,580	6.50E^7	9/9/03
26,041	219,340	12,638	6,762	6.60E^7	9/16/03
29,472	223,344	14,436	6,921	6.68E^7	9/23/03
24,920	226,032	11,911	7,039	6.77E^7	9/30/03
24,506	228,523	12,063	7,192	6.83E^7	10/7/03
31,506	231,740	14,882	7,403	6.94E^7	10/14/03
31,220	234,705	14,213	7,726	7.04E^7	10/21/03
27,692	237,381	12,664	7,939	7.14E^7	10/28/03
34,537	241,618	16,919	8,157	7.24E^7	11/4/03
40,424	246,140	18,809	8,377	7.37E^7	11/11/03
42,412	250,433	20,296	8,598	7.50E^7	11/18/03
38,330	254,261	18,054	8,841	7.62E^7	11/25/03
44,519	258,885	21,304	9,070	7.75E^7	12/2/03
51,604	264,439	24,236	9,348	7.91E^7	12/9/03
46,527	269,422	22,800	9,567	8.06E^7	12/16/03
24,030	272,055	13,545	9,678	8.13E^7	12/23/03
37,258	276,541	18,636	9,899	8.27E^7	12/30/03
38,373	280,877	19,006	10,151	8.40E^7	1/6/04
36,941	284,842	18,723	10,461	8.60E^7	1/13/04
40,118	289,194	19,638	10,699	8.73E^7	1/20/04
41,804	294,188	19,894	10,927	8.86E^7	1/27/04
50,840	299,796	23,466	11,356	9.03E^7	2/3/04

(continued)

Table A.1 (continued)

Weekly number of revisions	Cumulative number of articles revised	Weekly number of articles revised	Authors	Total cumulative delta (effort)	Week beginning date
54,238	305,522	24,729	11,704	$9.22E^7$	2/10/04
66,521	312,712	31,441	12,203	$9.38E^7$	2/17/04
69,213	319,493	31,161	12,665	$9.60E^7$	2/24/04
78,193	329,151	35,493	13,186	$9.83E^7$	3/2/04
79,887	338,293	36,811	13,719	$1.01E^8$	3/9/04
77,972	347,457	37,560	14,212	$1.04E^8$	3/16/04
77,155	357,159	36,345	14,731	$1.06E^8$	3/23/04
81,683	365,381	38,907	15,219	$1.08E^8$	3/30/04
73,003	373,381	35,376	15,680	$1.11E^8$	4/6/04
75,427	381,214	38,514	16,159	$1.13E^8$	4/13/04
74,138	388,869	34,676	16,709	$1.16E^8$	4/20/04
68,692	396,492	32,982	17,172	$1.18E^8$	4/27/04
71,759	403,411	34,478	17,667	$1.21E^8$	5/4/04
75,917	410,858	35,850	18,181	$1.23E^8$	5/11/04
78,402	419,953	37,882	18,704	$1.25E^8$	5/18/04
75,346	426,915	39,692	19,199	$1.27E^8$	5/25/04
118,349	432,570	72,964	19,682	$1.29E^8$	6/1/04
99,702	440,437	51,254	20,198	$1.31E^8$	6/8/04
96,729	448,271	51,700	20,631	$1.33E^8$	6/15/04
111,777	459,309	58,480	21,183	$1.36E^8$	6/22/04
106,100	473,845	57,989	21,718	$1.39E^8$	6/29/04
119,396	493,121	68,618	22,417	$1.41E^8$	7/6/04
130,251	514,680	73,713	23,113	$1.44E^8$	7/13/04
100,526	523,326	48,978	23,691	$1.47E^8$	7/20/04
87,462	530,930	44,332	24,336	$1.49E^8$	7/27/04
110,466	537,892	65,339	24,925	$1.52E^8$	8/3/04
113,967	546,290	57,879	25,551	$1.54E^8$	8/10/04
120,304	555,820	62,187	26,284	$1.57E^8$	8/17/04
120,058	565,246	58,461	26,947	$1.61E^8$	8/24/04
115,834	574,000	56,309	27,734	$1.64E^8$	8/31/04
124,769	583,059	59,178	28,515	$1.67E^8$	9/7/04
122,267	592,336	57,720	29,292	$1.71E^8$	9/14/04
132,127	602,240	66,816	30,148	$1.76E^8$	9/21/04
111,314	610,141	57,170	30,917	$1.79E^8$	9/28/04
124,746	618,689	62,832	31,848	$1.82E^8$	10/5/04
144,940	629,188	74,599	32,774	$1.87E^8$	10/12/04
136,101	639,428	65,410	33,642	$1.91E^8$	10/19/04
132,672	649,740	65,301	34,650	$1.96E^8$	10/26/04
152,128	659,013	70,838	35,583	$2.00E^8$	11/2/04
213,674	670,192	106,150	36,823	$2.07E^8$	11/9/04

(continued)

Table A.1 (continued)

Weekly number of revisions	Cumulative number of articles revised	Weekly number of articles revised	Authors	Total cumulative delta (effort)	Week beginning date
185,846	681,991	91,490	37,984	2.13E^8	11/16/04
156,254	693,497	71,871	38,976	2.18E^8	11/23/04
167,064	705,189	75,540	40,085	2.24E^8	11/30/04
150,252	716,136	71,121	41,140	2.30E^8	12/7/04
164,479	729,734	78,599	42,150	2.35E^8	12/14/04
162,469	742,587	84,757	43,102	2.41E^8	12/21/04
161,632	753,440	72,141	44,325	2.47E^8	12/28/04
128,400	761,984	57,392	45,388	2.54E^8	1/4/05
122,536	771,026	57,785	46,387	2.64E^8	1/11/05
136,179	780,517	62,418	47,454	2.73E^8	1/18/05
161,846	791,402	71,561	48,738	2.82E^8	1/25/05
153,226	801,552	67,517	49,975	2.90E^8	2/1/05
163,036	811,684	72,138	51,239	2.99E^8	2/8/05
172,671	822,719	76,581	52,554	3.08E^8	2/15/05
113,661	829,996	54,847	53,629	3.14E^8	2/22/05
161,854	840,545	75,643	54,983	3.24E^8	3/1/05
180,480	852,459	82,729	56,551	3.33E^8	3/8/05
171,234	863,650	78,476	57,910	3.43E^8	3/15/05
186,916	876,271	83,083	59,394	3.53E^8	3/22/05
201,589	890,357	89,754	60,950	3.64E^8	3/29/05
221,871	904,198	93,981	62,637	3.77E^8	4/5/05
239,502	920,185	103,877	64,355	3.89E^8	4/12/05
243,517	936,308	100,777	66,312	4.04E^8	4/19/05
263,798	951,895	119,988	68,104	4.21E^8	4/26/05
234,719	966,649	102,885	69,880	4.37E^8	5/3/05
241,134	979,892	103,136	71,665	4.58E^8	5/10/05
238,317	994,366	99,786	73,420	4.69E^8	5/17/05
243,028	1,008,350	103,132	75,267	4.79E^8	5/24/05
266,210	1,025,724	111,969	77,295	4.88E^8	5/31/05
245,884	1,042,880	104,679	79,097	4.95E^8	6/7/05
286,711	1,061,669	119,570	81,127	5.04E^8	6/14/05
266,907	1,077,812	109,882	83,136	5.13E^8	6/21/05
295,212	1,100,858	124,960	85,253	5.22E^8	6/28/05
320,991	1,119,864	125,910	87,793	5.34E^8	7/5/05
321,675	1,139,236	127,523	90,411	5.44E^8	7/12/05
329,430	1,158,846	131,161	93,244	5.54E^8	7/19/05
312,442	1,178,996	127,819	95,879	5.65E^8	7/26/05
328,770	1,200,433	133,097	98,875	5.78E^8	8/2/05
346,863	1,222,563	141,040	101,791	5.88E^8	8/9/05
326,891	1,242,653	135,631	104,766	6.00E^8	8/16/05

(continued)

Table A.1 (continued)

Weekly number of revisions	Cumulative number of articles revised	Weekly number of articles revised	Authors	Total cumulative delta (effort)	Week beginning date
346,848	1,263,040	140,283	107,395	6.12E^8	8/23/05
350,817	1,283,455	141,701	109,921	6.24E^8	8/30/05
317,127	1,300,880	132,722	112,427	6.37E^8	9/6/05
286,640	1,315,940	127,291	114,725	6.48E^8	9/13/05
362,041	1,333,421	152,579	117,661	6.65E^8	9/20/05
376,004	1,352,433	152,513	120,631	6.78E^8	9/27/05
371,650	1,370,957	151,810	123,716	6.93E^8	10/4/05
365,257	1,389,761	140,110	126,984	7.09E^8	10/11/05
409,742	1,409,093	159,402	130,304	7.28E^8	10/18/05
418,101	1,429,643	162,427	133,537	7.48E^8	10/25/05
438,250	1,449,812	176,803	137,282	7.68E^8	11/1/05
459,680	1,469,064	185,057	141,026	7.89E^8	11/8/05
409,954	1,488,055	161,334	144,407	8.08E^8	11/15/05
398,825	1,507,716	152,874	147,684	8.28E^8	11/22/05
496,192	1,529,766	176,414	151,952	8.52E^8	11/29/05
553,913	1,549,171	186,263	158,690	8.84E^8	12/6/05
603,482	1,571,499	201,928	166,235	9.22E^8	12/13/05
521,662	1,591,454	196,248	171,686	9.50E^8	12/20/05
555,333	1,615,067	205,132	177,321	9.74E^8	12/27/05
587,131	1,640,019	209,174	183,223	1.00E^9	1/3/06
600,123	1,664,451	209,325	189,514	1.06E^9	1/10/06
623,185	1,688,562	212,850	196,278	1.09E^9	1/17/06
665,231	1,712,250	234,042	202,857	1.11E^9	1/24/06
726,540	1,735,792	264,887	210,090	1.14E^9	1/31/06
621,844	1,759,469	213,175	218,219	1.18E^9	2/7/06
633,798	1,784,314	222,007	226,660	1.21E^9	2/14/06
649,760	1,807,059	229,633	235,229	1.24E^9	2/21/06
650,909	1,831,796	227,644	244,157	1.27E^9	2/28/06
699,865	1,856,640	254,835	253,075	1.31E^9	3/7/06
726,846	1,879,860	275,948	261,882	1.34E^9	3/14/06
691,099	1,903,440	243,380	270,919	1.37E^9	3/21/06
694,869	1,928,321	245,474	279,889	1.40E^9	3/28/06
690,346	1,951,924	248,630	288,460	1.42E^9	4/4/06
676,366	1,976,773	234,813	297,624	1.45E^9	4/11/06
686,915	2,001,391	226,081	307,431	1.48E^9	4/18/06
729,830	2,028,335	239,813	317,364	1.51E^9	4/25/06
723,374	2,054,631	235,435	327,319	1.54E^9	5/2/06
775,109	2,081,227	252,116	337,179	1.58E^9	5/9/06
785,343	2,106,953	255,331	347,115	1.61E^9	5/16/06
782,011	2,133,027	269,971	356,788	1.64E^9	5/23/06

(continued)

Table A.1 (continued)

Weekly number of revisions	Cumulative number of articles revised	Weekly number of articles revised	Authors	Total cumulative delta (effort)	Week beginning date
800,566	2,161,319	267,869	366,613	$1.67E^9$	5/30/06
770,135	2,188,901	260,302	376,157	$1.70E^9$	6/6/06
794,144	2,217,180	270,513	387,252	$1.73E^9$	6/13/06
783,549	2,244,143	263,942	397,867	$1.76E^9$	6/20/06
759,101	2,274,553	258,210	407,854	$1.78E^9$	6/27/06
768,596	2,303,109	260,848	418,321	$1.81E^9$	7/4/06
792,761	2,331,277	272,946	429,082	$1.84E^9$	7/11/06
821,173	2,358,609	290,224	439,816	$1.86E^9$	7/18/06
820,389	2,388,345	280,284	451,463	$1.89E^9$	7/25/06
842,902	2,417,419	283,616	464,656	$1.92E^9$	8/1/06
857,584	2,446,865	289,787	476,839	$1.95E^9$	8/8/06
852,343	2,477,628	304,437	488,713	$1.97E^9$	8/15/06
897,119	2,509,607	323,418	500,878	$2.00E^9$	8/22/06
868,988	2,538,234	292,130	513,527	$2.03E^9$	8/29/06
849,505	2,566,997	283,019	525,932	$2.07E^9$	9/5/06
844,226	2,590,777	283,507	538,267	$2.10E^9$	9/12/06
862,660	2,614,994	292,342	550,659	$2.14E^9$	9/19/06
885,715	2,646,867	306,258	562,907	$2.18E^9$	9/26/06
895,231	2,673,785	313,048	575,259	$2.23E^9$	10/3/06
893,133	2,700,502	286,542	588,141	$2.28E^9$	10/10/06
915,039	2,725,708	292,731	601,380	$2.32E^9$	10/17/06
918,532	2,751,473	282,590	615,142	$2.36E^9$	10/24/06
931,395	2,776,262	294,880	629,323	$2.41E^9$	10/31/06
970,212	2,804,103	306,533	643,966	$2.46E^9$	11/7/06
963,173	2,832,738	301,065	659,610	$2.50E^9$	11/14/06
933,700	2,860,489	307,611	673,795	$2.55E^9$	11/21/06
1,016,026	2,886,509	324,431	689,712	$2.59E^9$	11/28/06
1,033,755	2,917,898	329,522	706,257	$2.64E^9$	12/5/06
973,507	2,945,358	315,715	721,588	$2.68E^9$	12/12/06
811,895	2,968,454	275,750	733,683	$2.71E^9$	12/19/06
875,465	2,993,374	313,549	745,686	$2.74E^9$	12/26/06
1,031,674	3,019,473	339,357	761,163	$2.78E^9$	1/2/07
1,003,208	3,045,380	324,670	776,444	$2.82E^9$	1/9/07
1,007,007	3,072,686	321,547	792,234	$2.86E^9$	1/16/07
1,040,245	3,099,523	325,018	808,909	$2.91E^9$	1/23/07
1,069,685	3,127,198	330,837	827,190	$2.97E^9$	1/30/07
1,139,520	3,157,856	354,839	845,527	$3.02E^9$	2/6/07
1,113,174	3,185,500	349,049	863,167	$3.07E^9$	2/13/07
1,099,936	3,213,281	338,996	881,532	$3.13E^9$	2/20/07
1,086,941	3,238,930	341,060	899,376	$3.19E^9$	2/27/07

(continued)

Table A.1 (continued)

Weekly number of revisions	Cumulative number of articles revised	Weekly number of articles revised	Authors	Total cumulative delta (effort)	Week beginning date
1,095,356	3,265,411	337,817	916,925	3.24E^9	3/6/07
1,093,568	3,291,194	335,756	933,913	3.29E^9	3/13/07
1,084,710	3,317,351	328,254	951,607	3.35E^9	3/20/07
1,083,526	3,342,861	336,059	968,155	3.40E^9	3/27/07
1,065,959	3,368,317	340,749	984,028	3.44E^9	4/3/07
1,136,932	3,394,332	355,075	1,001,951	3.49E^9	4/10/07
1,109,833	3,420,278	345,491	1,019,196	3.55E^9	4/17/07
1,112,171	3,447,092	342,052	1,036,275	3.61E^9	4/24/07
1,067,388	3,470,758	326,105	1,052,343	3.66E^9	5/1/07
1,070,209	3,495,580	336,642	1,067,771	3.72E^9	5/8/07
1,051,737	3,521,820	328,410	1,082,759	3.77E^9	5/15/07
1,034,918	3,549,204	338,807	1,097,011	3.81E^9	5/22/07
1,070,679	3,577,659	370,146	1,110,935	3.86E^9	5/29/07
991,163	3,608,359	334,909	1,124,228	3.90E^9	6/5/07
968,328	3,643,284	336,670	1,136,640	3.93E^9	6/12/07
947,846	3,668,649	326,063	1,148,858	3.97E^9	6/19/07
944,726	3,693,514	319,996	1,161,571	4.00E^9	6/26/07
939,433	3,720,424	313,880	1,174,277	4.03E^9	7/3/07
954,016	3,753,369	326,199	1,187,463	4.06E^9	7/10/07
941,048	3,788,225	327,520	1,200,180	4.10E^9	7/17/07
950,973	3,830,355	336,995	1,213,054	4.13E^9	7/24/07
938,842	3,859,297	337,615	1,225,250	4.16E^9	7/31/07
884,209	3,888,248	312,593	1,237,419	4.18E^9	8/7/07
967,378	3,915,874	354,057	1,250,582	4.22E^9	8/14/07
926,836	3,947,775	332,712	1,263,597	4.25E^9	8/21/07
936,307	3,971,604	336,848	1,275,988	4.28E^9	8/28/07
961,612	4,005,934	337,242	1,289,337	4.32E^9	9/4/07
990,932	4,030,700	346,918	1,303,276	4.36E^9	9/11/07
993,041	4,054,109	340,612	1,317,240	4.40E^9	9/18/07
987,826	4,077,491	340,722	1,331,134	4.44E^9	9/25/07
1,003,507	4,106,923	354,966	1,345,258	4.49E^9	10/2/07
1,086,456	4,131,132	403,045	1,359,607	4.54E^9	10/9/07
999,954	4,156,566	331,205	1,374,487	4.58E^9	10/16/07
966,903	4,179,494	324,772	1,388,929	4.62E^9	10/23/07
962,905	4,209,699	329,033	1,403,051	4.66E^9	10/30/07
964,147	4,237,730	328,363	1,417,361	4.70E^9	11/6/07
1,002,126	4,260,757	361,031	1,431,429	4.75E^9	11/13/07
911,803	4,301,791	326,206	1,443,982	4.78E^9	11/20/07
1,048,460	4,325,541	419,051	1,457,699	4.82E^9	11/27/07
960,752	4,347,014	347,471	1,470,898	4.86E^9	12/4/07

(continued)

Table A.1 (continued)

Weekly number of revisions	Cumulative number of articles revised	Weekly number of articles revised	Authors	Total cumulative delta (effort)	Week beginning date
989,130	4,371,565	382,048	1,483,693	$4.90E^9$	12/11/07
902,534	4,408,972	368,307	1,493,962	$4.94E^9$	12/18/07
799,254	4,441,215	313,043	1,503,606	$4.96E^9$	12/25/07
918,685	4,478,009	327,451	1,516,027	$5.00E^9$	1/1/08
981,702	4,504,401	336,739	1,529,824	$5.04E^9$	1/8/08
994,130	4,532,950	350,118	1,543,757	$5.09E^9$	1/15/08
984,931	4,560,963	328,019	1,558,174	$5.13E^9$	1/22/08
1,007,180	4,593,547	345,374	1,572,482	$5.17E^9$	1/29/08
1,025,730	4,618,339	354,995	1,586,990	$5.22E^9$	2/5/08
1,000,222	4,647,328	338,923	1,601,470	$5.26E^9$	2/12/08
1,025,761	4,676,253	333,041	1,616,674	$5.31E^9$	2/19/08
1,040,687	4,701,652	358,040	1,632,103	$5.36E^9$	2/26/08
1,059,725	4,734,003	370,102	1,647,311	$5.40E^9$	3/4/08
1,043,632	4,780,635	374,444	1,661,953	$5.45E^9$	3/11/08
1,129,820	4,883,003	462,298	1,675,963	$5.49E^9$	3/18/08
1,132,339	4,929,173	427,666	1,690,684	$5.53E^9$	3/25/08
1,076,320	4,953,696	385,320	1,705,194	$5.58E^9$	4/1/08
1,012,179	4,977,485	339,678	1,719,945	$5.64E^9$	4/8/08
978,001	5,000,736	325,337	1,734,285	$5.68E^9$	4/15/08
1,007,853	5,024,747	340,861	1,748,420	$5.72E^9$	4/22/08
995,296	5,047,009	336,899	1,762,305	$5.77E^9$	4/29/08
977,519	5,069,289	336,900	1,776,088	$5.82E^9$	5/6/08
999,181	5,092,398	347,122	1,789,874	$5.86E^9$	5/13/08
972,592	5,115,392	344,079	1,802,661	$5.90E^9$	5/20/08
1,005,059	5,139,484	363,377	1,815,526	$5.94E^9$	5/27/08
984,435	5,163,512	364,125	1,827,887	$5.98E^9$	6/3/08
981,901	5,188,680	375,468	1,839,775	$6.01E^9$	6/10/08
952,702	5,213,809	362,562	1,850,790	$6.04E^9$	6/17/08
897,344	5,238,377	331,017	1,861,520	$6.06E^9$	6/24/08
933,901	5,262,720	353,564	1,872,036	$6.09E^9$	7/1/08
944,742	5,286,932	366,044	1,882,841	$6.12E^9$	7/8/08
915,842	5,311,676	343,063	1,893,614	$6.15E^9$	7/15/08
909,266	5,334,534	338,610	1,904,488	$6.18E^9$	7/22/08
899,251	5,361,291	333,527	1,915,883	$6.20E^9$	7/29/08
934,258	5,387,962	348,148	1,927,133	$6.23E^9$	8/5/08
968,280	5,415,711	373,311	1,938,053	$6.27E^9$	8/12/08
968,643	5,448,782	367,358	1,948,968	$6.30E^9$	8/19/08
964,894	5,474,533	366,888	1,960,154	$6.33E^9$	8/26/08
966,342	5,494,773	361,516	1,971,996	$6.37E^9$	9/2/08
993,713	5,514,485	370,615	1,984,121	$6.41E^9$	9/9/08

(continued)

Table A.1 (continued)

Weekly number of revisions	Cumulative number of articles revised	Weekly number of articles revised	Authors	Total cumulative delta (effort)	Week beginning date
1,111,992	5,535,319	475,518	1,996,060	6.45E^9	9/16/08
1,049,316	5,557,879	420,501	2,008,422	6.50E^9	9/23/08
1,119,758	5,585,788	471,446	2,020,913	6.54E^9	9/30/08
1,064,504	5,632,285	427,095	2,033,820	6.59E^9	10/7/08
1,010,482	5,656,807	375,797	2,046,923	6.63E^9	10/14/08
992,502	5,680,965	355,782	2,059,962	6.67E^9	10/21/08
968,838	5,701,989	347,471	2,072,829	6.71E^9	10/28/08
1,001,765	5,721,265	399,874	2,085,700	6.75E^9	11/4/08
984,570	5,743,413	366,814	2,098,838	6.79E^9	11/11/08
945,368	5,763,351	344,070	2,111,630	6.83E^9	11/18/08
903,356	5,782,971	347,950	2,123,340	6.87E^9	11/25/08
928,210	5,802,759	329,309	2,136,501	6.91E^9	12/2/08
993,492	5,825,536	383,136	2,149,134	6.94E^9	12/9/08
909,620	5,846,377	354,744	2,159,907	6.98E^9	12/16/08
764,646	5,864,445	308,661	2,168,900	7.00E^9	12/23/08
975,283	5,884,892	406,783	2,179,926	7.03E^9	12/30/08
983,916	5,909,301	365,022	2,192,520	7.07E^9	1/6/09
969,564	5,933,155	349,114	2,205,427	7.11E^9	1/13/09
974,225	5,955,773	340,006	2,218,561	7.15E^9	1/20/09
1,007,728	5,977,291	354,269	2,232,045	7.19E^9	1/27/09
1,015,169	5,999,559	360,504	2,245,812	7.23E^9	2/3/09
1,044,344	6,024,384	402,979	2,259,539	7.27E^9	2/10/09
1,103,116	6,050,182	429,451	2,273,453	7.31E^9	2/17/09
1,054,590	6,074,186	368,725	2,287,501	7.36E^9	2/24/09
1,018,180	6,104,691	359,950	2,301,075	7.40E^9	3/3/09
1,009,373	6,129,325	361,421	2,314,582	7.44E^9	3/10/09
957,155	6,153,165	344,460	2,327,398	7.47E^9	3/17/09
964,197	6,180,657	363,723	2,340,121	7.49E^9	3/24/09
964,664	6,209,744	368,802	2,352,369	7.51E^9	3/31/09
946,688	6,230,917	353,119	2,364,711	7.53E^9	4/7/09
942,987	6,260,974	349,220	2,377,033	7.56E^9	4/14/09
930,019	6,284,667	338,139	2,390,149	7.58E^9	4/21/09
936,025	6,305,621	346,839	2,402,804	7.61E^9	4/28/09
937,017	6,326,376	348,757	2,415,485	7.63E^9	5/5/09
978,798	6,347,205	385,713	2,428,519	7.65E^9	5/12/09
1,001,224	6,369,465	413,511	2,440,355	7.67E^9	5/19/09
993,787	6,391,774	392,927	2,452,870	7.70E^9	5/26/09
935,155	6,413,748	354,403	2,465,139	7.72E^9	6/2/09
922,752	6,437,626	358,340	2,477,053	7.74E^9	6/9/09
909,640	6,468,282	355,032	2,488,798	7.76E^9	6/16/09

(continued)

Table A.1 (continued)

Weekly number of revisions	Cumulative number of articles revised	Weekly number of articles revised	Authors	Total cumulative delta (effort)	Week beginning date
924,520	6,516,616	380,946	2,500,338	$7.78E^9$	6/23/09
853,266	6,538,632	337,657	2,511,254	$7.80E^9$	6/30/09
846,162	6,565,374	329,239	2,522,454	$7.82E^9$	7/7/09
870,037	6,590,070	341,012	2,533,758	$7.84E^9$	7/14/09
880,154	6,611,993	347,514	2,545,287	$7.86E^9$	7/21/09
850,054	6,632,516	321,808	2,555,965	$7.87E^9$	7/28/09
903,246	6,654,068	370,548	2,566,807	$7.89E^9$	8/4/09
887,747	6,680,801	357,102	2,578,035	$7.91E^9$	8/11/09
876,585	6,703,216	344,917	2,588,776	$7.93E^9$	8/18/09
889,818	6,726,778	349,122	2,600,464	$7.95E^9$	8/25/09
887,896	6,745,321	356,964	2,611,634	$7.97E^9$	9/1/09
948,206	6,763,553	395,567	2,622,982	$7.99E^9$	9/8/09
951,367	6,782,839	395,521	2,634,769	$8.02E^9$	9/15/09
939,898	6,802,346	365,129	2,646,483	$8.04E^9$	9/22/09
924,395	6,822,168	367,021	2,657,922	$8.06E^9$	9/29/09
892,537	6,840,604	331,956	2,670,171	$8.08E^9$	10/6/09
929,323	6,860,348	364,447	2,682,560	$8.11E^9$	10/13/09
886,441	6,879,637	331,649	2,694,986	$8.13E^9$	10/20/09
910,818	6,898,859	352,653	2,707,369	$8.15E^9$	10/27/09
924,563	6,918,432	349,690	2,720,062	$8.18E^9$	11/3/09
885,810	6,935,707	341,624	2,732,663	$8.20E^9$	11/10/09
906,688	6,954,758	357,452	2,745,060	$8.22E^9$	11/17/09
948,276	6,973,509	415,695	2,757,041	$8.24E^9$	11/24/09
980,905	6,992,764	422,914	2,769,483	$8.26E^9$	12/1/09
935,833	7,011,318	406,145	2,781,736	$8.29E^9$	12/8/09
1,016,347	7,031,534	505,471	2,792,523	$8.31E^9$	12/15/09
1,117,004	7,050,244	638,105	2,801,521	$8.32E^9$	12/22/09
837,587	7,077,862	338,384	2,812,137	$8.34E^9$	12/29/09
918,229	7,099,847	344,245	2,824,692	$8.37E^9$	1/5/10
907,482	7,118,981	338,168	2,836,857	$8.39E^9$	1/12/10
911,048	7,141,265	339,050	2,849,315	$8.41E^9$	1/19/10
898,988	7,160,595	325,647	2,862,440	$8.43E^9$	1/26/10
593,445	7,164,222	170,744	2,878,127	$8.44E^9$	2/2/10
566,743	7,166,718	162,589	2,891,338	$8.45E^9$	2/9/10
596,067	7,169,127	162,622	2,904,573	$8.46E^9$	2/16/10
583,422	7,171,513	159,011	2,917,089	$8.47E^9$	2/23/10
582,386	7,173,937	163,806	2,929,032	$8.48E^9$	3/2/10
585,733	7,176,688	170,930	2,940,590	$8.49E^9$	3/9/10
557,328	7,178,828	157,064	2,951,543	$8.49E^9$	3/16/10
564,313	7,181,545	158,041	2,962,319	$8.50E^9$	3/23/10

(continued)

Table A.1 (continued)

Weekly number of revisions	Cumulative number of articles revised	Weekly number of articles revised	Authors	Total cumulative delta (effort)	Week beginning date
559,254	7,183,789	164,378	2,972,568	8.51E^9	3/30/10
591,199	7,186,360	169,873	2,983,248	8.52E^9	4/6/10
601,953	7,189,960	171,943	2,994,048	8.53E^9	4/13/10
598,252	7,192,848	168,767	3,004,940	8.53E^9	4/20/10
592,159	7,195,487	168,322	3,015,784	8.54E^9	4/27/10
611,872	7,198,870	173,142	3,026,259	8.55E^9	5/4/10
604,196	7,202,151	169,137	3,037,059	8.56E^9	5/11/10
604,065	7,205,860	181,544	3,047,086	8.56E^9	5/18/10
737,295	7,227,243	264,043	3,058,778	8.57E^9	5/25/10
744,679	7,238,967	263,425	3,070,256	8.58E^9	6/1/10
842,388	7,254,137	348,664	3,082,199	8.58E^9	6/8/10
756,629	7,270,373	295,069	3,093,221	8.59E^9	6/15/10
618,492	7,286,273	196,094	3,103,177	8.60E^9	6/22/10
536,219	7,295,060	176,854	3,111,021	8.60E^9	6/29/10
71,602	7,296,360	32,020	3,112,211	8.60E^9	7/6/10

infrastructure managed by the Rosen Center for Advanced Computing (Purdue University 2008). This system was powerful enough to handle most of the computational load of this project.

Similarly, most standard software packages for network analysis would have been unable to handle such a large quantity of data or would have taken many years of processing to complete the analysis. Consequently, the second author of this volume and two other programmers, David Braun and Dennis Lazar of the Rosen Center for Advanced Computing, designed and wrote several custom Java programs. These programs moved the data from the raw text file into a SQL database to be queried; constructed network graphs from that database as described above; calculated the inbound degree centralization, outbound degree centralization, betweenness centralization, and assortativity from each of those network graphs; and calculated the social entropy from the SQL database for the same weeklong time ranges.

This collaborative programming effort was undertaken as part of the Visible Effort/KredibleNet research initiative, serving the needs of multiple research studies (Matei et al. 2015), including the present analysis. As a whole, after the first 3 weeks and the final week of revisions were removed from the data set (see above), the computational process conducted by Britt, Braun, and Lazar distilled the 67 GB text file into a core data set of five variables, each of which was evaluated 491 times, corresponding with the 491 weeks of Wikipedia's growth from February 6, 2001, to July 5, 2010.

Table A.2 Weekly values of five key measures of Wikipedia activity

Week start date	Week number	Assort.	Out. degree cent.	Inbound degree cent.	Between. cent.	Entropy
1/16/01	1	Not est.	Not est.	Not est.	Not est.	0.4690
1/23/01	2	Not est.	Not est.	Not est.	Not est.	0.7517
1/30/01	3	Not est.	Not est.	Not est.	Not est.	0.7588
2/6/01	4	−0.4759	0.2831	0.5394	0.2500	0.7940
2/13/01	5	0.4976	0.8105	0.8105	0.0833	0.7780
2/20/01	6	−0.3521	0.7520	0.8707	0.2351	0.7104
2/27/01	7	−0.5149	0.9739	0.9852	0.3333	0.6865
3/6/01	8	−0.5175	0.3436	0.9460	0.2469	0.7312
3/13/01	9	−0.7659	0.5807	0.4129	0.2381	0.6898
3/20/01	10	−0.2492	0.9554	0.9485	0.0417	0.7610
3/27/01	11	−0.0024	0.9683	0.9673	0.0781	0.7623
4/3/01	12	0.2540	0.7773	0.9477	0.0000	0.7771
4/10/01	13	−0.6811	0.9598	0.8452	0.3333	0.7650
4/17/01	14	0.6476	0.7983	0.9992	0.0000	0.7618
4/24/01	15	0.7217	0.9799	0.8209	0.2000	0.7477
5/1/01	16	0.0000	1.0000	1.0000	0.0000	0.7415
5/8/01	17	0.3258	0.9101	0.8917	0.0000	0.7427
5/15/01	18	0.0250	0.3707	0.3369	0.1736	0.7750
5/22/01	19	0.3628	0.9839	0.9839	0.3667	0.7682
5/29/01	20	−0.3980	0.3617	0.8136	0.1361	0.7600
6/5/01	21	0.2392	0.9326	0.9323	0.2315	0.7597
6/12/01	22	−0.2657	0.5764	0.8192	0.1543	0.7600
6/19/01	23	−0.6468	0.2486	0.3014	0.1871	0.7556
6/26/01	24	0.0666	0.6858	0.4414	0.1735	0.7666
7/3/01	25	−0.3711	0.1661	0.2598	0.0268	0.7632
7/10/01	26	−0.3744	0.3799	0.5371	0.1585	0.7707
7/17/01	27	−0.1817	0.9419	0.9229	0.0719	0.7665
7/24/01	28	−0.6593	0.8682	0.4812	0.2429	0.7452
7/31/01	29	−0.3860	0.5274	0.6218	0.2231	0.7551
8/7/01	30	−0.7357	0.5221	0.7261	0.3012	0.7672
8/14/01	31	−0.3980	0.7166	0.7429	0.3015	0.7647
8/21/01	32	−0.5167	0.6154	0.4221	0.2061	0.7648
8/28/01	33	−0.1199	0.9140	0.9594	0.2124	0.7736
9/4/01	34	−0.4317	0.5402	0.6945	0.3836	0.7722
9/11/01	35	0.2634	0.8559	0.5893	0.3001	0.7734
9/18/01	36	−0.7840	0.2300	0.4419	0.2686	0.7653
9/25/01	37	−0.4912	0.2606	0.5304	0.3705	0.7839
10/2/01	38	−0.5190	0.5423	0.8527	0.2357	0.7869
10/9/01	39	−0.3552	0.6739	0.3608	0.1958	0.7897
10/16/01	40	−0.4516	0.3487	0.4435	0.3444	0.7916
10/23/01	41	−0.3914	0.3419	0.3418	0.3224	0.7895

(continued)

Table A.2 (continued)

Week start date	Week number	Assort.	Out. degree cent.	Inbound degree cent.	Between. cent.	Entropy
10/30/01	42	0.2709	0.7911	0.7911	0.3087	0.7968
11/6/01	43	−0.3796	0.5132	0.7051	0.2809	0.7947
11/13/01	44	−0.2819	0.3425	0.2981	0.4587	0.7948
11/20/01	45	−0.2922	0.4613	0.1768	0.4494	0.7925
11/27/01	46	−0.3294	0.5928	0.4411	0.4734	0.7862
12/4/01	47	−0.2015	0.4103	0.4149	0.4331	0.7767
12/11/01	48	−0.2957	0.2544	0.2773	0.4914	0.7784
12/18/01	49	−0.3148	0.2411	0.3745	0.3886	0.7788
12/25/01	50	−0.2514	0.4635	0.4927	0.3640	0.7768
1/1/02	51	−0.3125	0.3054	0.4887	0.3799	0.7755
1/8/02	52	−0.2983	0.2913	0.3442	0.4092	0.7732
1/15/02	53	−0.2899	0.2536	0.2365	0.4483	0.7698
1/22/02	54	−0.4040	0.7573	0.1555	0.3636	0.7679
1/29/02	55	−0.4994	0.9321	0.3167	0.2850	0.7649
2/5/02	56	−0.4397	0.7604	0.4376	0.3252	0.7612
2/12/02	57	−0.4065	0.8541	0.1418	0.3456	0.7592
2/19/02	58	−0.5514	0.8663	0.1902	0.3320	0.7498
2/26/02	59	−0.3499	0.1042	0.7514	0.3726	0.7485
3/5/02	60	−0.2812	0.1673	0.7834	0.3045	0.7477
3/12/02	61	−0.3269	0.2121	0.7781	0.3378	0.7456
3/19/02	62	−0.2746	0.4569	0.6557	0.3244	0.7454
3/26/02	63	−0.2981	0.2584	0.6404	0.2807	0.7454
4/2/02	64	−0.2384	0.4743	0.7701	0.2371	0.7441
4/9/02	65	−0.0467	0.9272	0.7461	0.2616	0.7374
4/16/02	66	−0.2861	0.3457	0.8110	0.2458	0.7375
4/23/02	67	−0.2864	0.4412	0.7617	0.2368	0.7372
4/30/02	68	−0.2771	0.2370	0.5398	0.2366	0.7369
5/7/02	69	−0.3347	0.9760	0.9697	0.2252	0.7364
5/14/02	70	0.0026	0.8769	0.8633	0.2122	0.7357
5/21/02	71	−0.0587	0.9623	0.8456	0.2088	0.7351
5/28/02	72	−0.2409	0.3093	0.6791	0.2026	0.7336
6/4/02	73	−0.2647	0.1500	0.5778	0.2440	0.7337
6/11/02	74	−0.2200	0.6637	0.8319	0.2631	0.7334
6/18/02	75	−0.2600	0.4644	0.3733	0.2368	0.7338
6/25/02	76	−0.2493	0.5671	0.6437	0.1715	0.7322
7/2/02	77	−0.2338	0.5465	0.8810	0.1880	0.7287
7/9/02	78	−0.0291	0.7004	0.4659	0.2177	0.7277
7/16/02	79	−0.2554	0.2387	0.4509	0.2229	0.7284
7/23/02	80	−0.2232	0.2719	0.4801	0.2070	0.7271
7/30/02	81	−0.1863	0.3922	0.4480	0.2516	0.7247
8/6/02	82	−0.0342	0.8211	0.7770	0.2315	0.7253

(continued)

Table A.2 (continued)

Week start date	Week number	Assort.	Out. degree cent.	Inbound degree cent.	Between. cent.	Entropy
8/13/02	83	−0.0366	0.9002	0.8121	0.2235	0.7247
8/20/02	84	−0.2013	0.3423	0.8895	0.2035	0.7243
8/27/02	85	−0.1797	0.1675	0.6147	0.2072	0.7233
9/3/02	86	−0.1831	0.1240	0.4731	0.2186	0.7236
9/10/02	87	−0.1577	0.3030	0.5378	0.2096	0.7225
9/17/02	88	−0.1611	0.2035	0.4515	0.2014	0.7186
9/24/02	89	−0.1875	0.2812	0.6539	0.2292	0.7012
10/1/02	90	−0.1818	0.3879	0.4546	0.2257	0.6975
10/8/02	91	−0.1800	0.2686	0.7622	0.2082	0.6966
10/15/02	92	−0.1522	0.1660	0.4131	0.1778	0.5736
10/22/02	93	−0.1611	0.1990	0.4444	0.1936	0.4895
10/29/02	94	−0.1716	0.3691	0.7214	0.1925	0.4925
11/5/02	95	−0.1286	0.4839	0.3944	0.1846	0.4940
11/12/02	96	−0.1762	0.1640	0.4767	0.2121	0.4965
11/19/02	97	−0.1600	0.3069	0.5532	0.1871	0.4943
11/26/02	98	−0.1710	0.4466	0.4707	0.1920	0.4962
12/3/02	99	−0.1622	0.3225	0.5681	0.2087	0.4978
12/10/02	100	−0.1267	0.9444	0.9917	0.1960	0.5033
12/17/02	101	−0.1682	0.3750	0.8039	0.2027	0.5058
12/24/02	102	−0.0094	0.4052	0.4822	0.1922	0.5084
12/31/02	103	−0.0482	0.3799	0.4114	0.1944	0.5127
1/7/03	104	−0.1464	0.1360	0.3256	0.1826	0.5162
1/14/03	105	−0.1278	0.2484	0.5092	0.1840	0.5191
1/21/03	106	−0.1189	0.9092	0.9134	0.2192	0.5198
1/28/03	107	−0.1370	0.4618	0.7167	0.2125	0.5207
2/4/03	108	−0.0534	0.4737	0.3237	0.2189	0.5243
2/11/03	109	−0.1007	0.2606	0.4986	0.1876	0.5272
2/18/03	110	−0.1115	0.9973	0.9973	0.1796	0.5310
2/25/03	111	−0.0679	0.5584	0.4728	0.1584	0.5336
3/4/03	112	−0.0951	0.3049	0.6117	0.1836	0.5362
3/11/03	113	−0.1012	0.2378	0.5494	0.1641	0.5393
3/18/03	114	−0.1095	0.2499	0.5445	0.1661	0.5414
3/25/03	115	−0.1165	0.3006	0.6628	0.1742	0.5436
4/1/03	116	−0.1020	0.2424	0.6429	0.1489	0.5454
4/8/03	117	−0.1281	0.2115	0.6872	0.1716	0.5479
4/15/03	118	−0.1106	0.3389	0.5018	0.1616	0.5498
4/22/03	119	−0.1113	0.2829	0.6990	0.1475	0.5526
4/29/03	120	−0.1224	0.2787	0.6210	0.1491	0.5553
5/6/03	121	−0.1129	0.2173	0.4571	0.1529	0.5580
5/13/03	122	−0.0991	0.2354	0.6083	0.1599	0.5608
5/20/03	123	−0.1058	0.2695	0.5860	0.1681	0.5631

(continued)

Table A.2 (continued)

Week start date	Week number	Assort.	Out. degree cent.	Inbound degree cent.	Between. cent.	Entropy
5/27/03	124	−0.1114	0.2282	0.5178	0.1689	0.5649
6/3/03	125	−0.0987	0.3814	0.4331	0.1686	0.5672
6/10/03	126	−0.1083	0.4523	0.5566	0.1627	0.5691
6/17/03	127	−0.1098	0.2652	0.6215	0.1697	0.5707
6/24/03	128	−0.1100	0.1697	0.4749	0.1607	0.5727
7/1/03	129	−0.1015	0.3479	0.5039	0.1626	0.5691
7/8/03	130	−0.1192	0.5421	0.4467	0.1661	0.5702
7/15/03	131	−0.1151	0.7804	0.7124	0.1678	0.5711
7/22/03	132	−0.1068	0.4414	0.8332	0.1636	0.5722
7/29/03	133	−0.1076	0.2030	0.5112	0.1615	0.5735
8/5/03	134	−0.1109	0.1911	0.5537	0.1757	0.5754
8/12/03	135	−0.0917	0.2319	0.4769	0.1590	0.5786
8/19/03	136	−0.0984	0.3929	0.7957	0.1714	0.5809
8/26/03	137	−0.1033	0.1870	0.5138	0.1584	0.5836
9/2/03	138	−0.1016	0.2386	0.6735	0.1545	0.5852
9/9/03	139	−0.1024	0.2892	0.8464	0.1613	0.5871
9/16/03	140	−0.0868	0.3287	0.6625	0.1472	0.5891
9/23/03	141	−0.0984	0.2628	0.6360	0.1546	0.5909
9/30/03	142	−0.0589	0.9251	0.9515	0.1492	0.5924
10/7/03	143	−0.0796	0.2886	0.6710	0.1355	0.5936
10/14/03	144	−0.0860	0.4863	0.9055	0.1456	0.5948
10/21/03	145	−0.0785	0.8010	0.9092	0.1388	0.5958
10/28/03	146	−0.0957	0.4392	0.5746	0.1433	0.5976
11/4/03	147	−0.1021	0.5394	0.8938	0.1513	0.5993
11/11/03	148	−0.0982	0.1760	0.5514	0.1482	0.6014
11/18/03	149	−0.0988	0.1560	0.7242	0.1442	0.6034
11/25/03	150	−0.0997	0.3471	0.7528	0.1526	0.6053
12/2/03	151	−0.0869	0.4783	0.8437	0.1354	0.6075
12/9/03	152	−0.0981	0.2221	0.6806	0.1403	0.6100
12/16/03	153	−0.0962	0.9313	0.9940	0.1512	0.6119
12/23/03	154	−0.1031	0.3551	0.6510	0.1083	0.6128
12/30/03	155	−0.0959	0.2519	0.4632	0.1385	0.6149
1/6/04	156	−0.0892	0.2803	0.3034	0.1528	0.6167
1/13/04	157	−0.0977	0.3018	0.7406	0.1513	0.6174
1/20/04	158	−0.0891	0.1729	0.7614	0.1349	0.6190
1/27/04	159	−0.0954	0.1643	0.6136	0.1376	0.6209
2/3/04	160	−0.0979	0.1763	0.9180	0.1547	0.6220
2/10/04	161	−0.0917	0.1261	0.7994	0.1527	0.6237
2/17/04	162	−0.0906	0.3108	0.5910	0.1487	0.6250
2/24/04	163	−0.0795	0.2628	0.6674	0.1429	0.6273
3/2/04	164	−0.0859	0.2897	0.9072	0.1625	0.6298

(continued)

Table A.2 (continued)

Week start date	Week number	Assort.	Out. degree cent.	Inbound degree cent.	Between. cent.	Entropy
3/9/04	165	−0.0834	0.1710	0.6148	0.1668	0.6327
3/16/04	166	−0.0845	0.1526	0.5075	0.1681	0.6354
3/23/04	167	−0.0839	0.3011	0.5707	0.1624	0.6373
3/30/04	168	−0.0831	0.2200	0.7284	0.1547	0.6393
4/6/04	169	−0.0783	0.4206	0.6914	0.1601	0.6416
4/13/04	170	−0.0816	0.2005	0.7359	0.1508	0.6431
4/20/04	171	0.0014	0.3639	0.3638	0.1573	0.6453
4/27/04	172	−0.0755	0.3009	0.5679	0.1471	0.6472
5/4/04	173	−0.0754	0.1418	0.4358	0.1446	0.6490
5/11/04	174	−0.0780	0.1377	0.7009	0.1479	0.6506
5/18/04	175	−0.0747	0.3591	0.5729	0.1398	0.6520
5/25/04	176	−0.0735	0.2544	0.7273	0.1284	0.6531
6/1/04	177	−0.0971	0.2783	0.7566	0.1244	0.6542
6/8/04	178	−0.0797	0.2294	0.3803	0.1245	0.6556
6/15/04	179	−0.0770	0.2990	0.8946	0.1271	0.6571
6/22/04	180	−0.0736	0.3041	0.7347	0.1233	0.6588
6/29/04	181	−0.0607	0.1373	0.1461	0.1292	0.6604
7/6/04	182	−0.0741	0.1464	0.5771	0.1183	0.6614
7/13/04	183	−0.0694	0.3946	0.6740	0.1332	0.6629
7/20/04	184	−0.0699	0.3279	0.6863	0.1199	0.6644
7/27/04	185	−0.0746	0.2879	0.9212	0.1227	0.6659
8/3/04	186	−0.0737	0.2764	0.4490	0.1172	0.6673
8/10/04	187	−0.0683	0.1826	0.9364	0.1205	0.6689
8/17/04	188	−0.0699	0.1841	0.6792	0.1237	0.6703
8/24/04	189	−0.0698	0.8192	0.9611	0.1240	0.6719
8/31/04	190	−0.0717	0.2020	0.7668	0.1233	0.6730
9/7/04	191	−0.0669	0.1566	0.8375	0.0953	0.6743
9/14/04	192	−0.0705	0.2350	0.7053	0.1170	0.6754
9/21/04	193	0.0006	0.9937	0.9936	0.1217	0.6760
9/28/04	194	−0.0683	0.2498	0.8887	0.0994	0.6769
10/5/04	195	−0.0724	0.1306	0.6268	0.1076	0.6777
10/12/04	196	−0.0693	0.3028	0.4587	0.1151	0.6791
10/19/04	197	−0.0710	0.3965	0.8384	0.1269	0.6809
10/26/04	198	−0.0687	0.7466	0.9468	0.1252	0.6817
11/2/04	199	−0.0693	0.2194	0.7468	0.1215	0.6827
11/9/04	200	−0.0628	0.4990	0.4993	0.1289	0.6827
11/16/04	201	−0.0626	0.9588	0.9969	0.1305	0.6837
11/23/04	202	−0.0627	0.1537	0.7095	0.1106	0.6850
11/30/04	203	−0.0614	0.9721	0.9826	0.1248	0.6866
12/7/04	204	−0.0645	0.9832	0.9887	0.1151	0.6876
12/14/04	205	−0.0630	0.7514	0.9364	0.1209	0.6888

(continued)

Table A.2 (continued)

Week start date	Week number	Assort.	Out. degree cent.	Inbound degree cent.	Between. cent.	Entropy
12/21/04	206	−0.0619	0.7096	0.9322	0.1055	0.6899
12/28/04	207	−0.0625	0.3142	0.8228	0.1108	0.6916
1/4/05	208	−0.0624	0.1741	0.7898	0.1131	0.6934
1/11/05	209	−0.0621	0.1665	0.5516	0.1032	0.6968
1/18/05	210	−0.0617	0.2625	0.3770	0.1017	0.6988
1/25/05	211	−0.0623	0.2424	0.2973	0.1110	0.7007
2/1/05	212	−0.0600	0.3023	0.2641	0.1053	0.7022
2/8/05	213	−0.0591	0.3075	0.2481	0.1109	0.7037
2/15/05	214	−0.0581	0.2887	0.2017	0.1087	0.7050
2/22/05	215	−0.0618	0.2305	0.2309	0.0922	0.7059
3/1/05	216	−0.0591	0.2697	0.2185	0.1022	0.7070
3/8/05	217	−0.0560	0.3153	0.2622	0.1020	0.7081
3/15/05	218	−0.0579	0.3132	0.2461	0.0995	0.7091
3/22/05	219	−0.0587	0.3020	0.2830	0.1047	0.7103
3/29/05	220	−0.0573	0.2476	0.3030	0.1107	0.7120
4/5/05	221	−0.0556	0.1772	0.2624	0.1124	0.7133
4/12/05	222	−0.0529	0.2427	0.3151	0.1111	0.7147
4/19/05	223	−0.0528	0.3254	0.2364	0.1110	0.7160
4/26/05	224	−0.0493	0.1979	0.3022	0.0992	0.7166
5/3/05	225	−0.0544	0.3635	0.2220	0.0993	0.7174
5/10/05	226	−0.0534	0.3989	0.3053	0.0971	0.7194
5/17/05	227	−0.0495	0.5614	0.8029	0.1007	0.7190
5/24/05	228	−0.0488	0.9962	0.9992	0.1010	0.7181
5/31/05	229	−0.0487	0.9123	0.9654	0.1019	0.7175
6/7/05	230	−0.0503	0.3635	0.6217	0.0949	0.7174
6/14/05	231	−0.0491	0.1644	0.7397	0.0995	0.7171
6/21/05	232	−0.0504	0.1158	0.6304	0.0959	0.7166
6/28/05	233	−0.0508	0.1579	0.7644	0.0972	0.7165
7/5/05	234	−0.0467	0.2148	0.5445	0.0960	0.7164
7/12/05	235	−0.0480	0.4485	0.8047	0.0983	0.7163
7/19/05	236	−0.0494	0.1846	0.7392	0.0971	0.7161
7/26/05	237	−0.0489	0.3130	0.6096	0.0975	0.7160
8/2/05	238	−0.0469	0.2969	0.7614	0.0970	0.7155
8/9/05	239	−0.0460	0.2655	0.4546	0.0969	0.7155
8/16/05	240	−0.0470	0.2278	0.6530	0.0974	0.7151
8/23/05	241	−0.0463	0.1723	0.6270	0.0925	0.7150
8/30/05	242	−0.0432	0.9202	0.9761	0.0841	0.7142
9/6/05	243	−0.0459	0.3286	0.8460	0.0833	0.7136
9/13/05	244	−0.0458	0.6437	0.9739	0.0742	0.7130
9/20/05	245	−0.0440	0.9326	0.9686	0.0803	0.7112
9/27/05	246	−0.0401	0.5452	0.4681	0.0807	0.7108

(continued)

Table A.2 (continued)

Week start date	Week number	Assort.	Out. degree cent.	Inbound degree cent.	Between. cent.	Entropy
10/4/05	247	−0.0442	0.2365	0.7691	0.0850	0.7102
10/11/05	248	−0.0393	0.9647	0.9913	0.0858	0.7093
10/18/05	249	−0.0416	0.1314	0.5391	0.0847	0.7082
10/25/05	250	−0.0416	0.3635	0.8208	0.0860	0.7066
11/1/05	251	−0.0413	0.1943	0.7496	0.0841	0.7048
11/8/05	252	−0.0393	0.7130	0.9019	0.0810	0.7035
11/15/05	253	−0.0381	0.4567	0.8896	0.0777	0.7028
11/22/05	254	−0.0392	0.3928	0.7950	0.0755	0.7013
11/29/05	255	−0.0358	0.1728	0.7209	0.0839	0.7002
12/6/05	256	−0.0337	0.1641	0.7011	0.0919	0.6975
12/13/05	257	−0.0315	0.2318	0.7233	0.0955	0.6953
12/20/05	258	−0.0109	0.3736	0.8232	0.0854	0.6939
12/27/05	259	−0.0344	0.1533	0.6820	0.0878	0.6933
1/3/06	260	−0.0315	0.3301	0.7646	0.0870	0.6921
1/10/06	261	−0.0270	0.4035	0.4108	0.0903	0.6862
1/17/06	262	−0.0308	0.1839	0.6673	0.0910	0.6851
1/24/06	263	−0.0299	0.4680	0.7969	0.0878	0.6840
1/31/06	264	−0.0291	0.2777	0.3988	0.0834	0.6831
2/7/06	265	−0.0293	0.6569	0.9261	0.0856	0.6817
2/14/06	266	−0.0281	0.8035	0.8034	0.0830	0.6808
2/21/06	267	−0.0289	0.2407	0.6392	0.0826	0.6789
2/28/06	268	−0.0284	0.2065	0.6466	0.0806	0.6779
3/7/06	269	−0.0276	0.1754	0.7467	0.0755	0.6769
3/14/06	270	−0.0261	0.2663	0.7440	0.0630	0.6756
3/21/06	271	−0.0251	0.4643	0.8602	0.0667	0.6745
3/28/06	272	−0.0239	0.7368	0.9452	0.0734	0.6732
4/4/06	273	−0.0268	0.3142	0.7754	0.0693	0.6719
4/11/06	274	−0.0255	0.3620	0.6943	0.0696	0.6709
4/18/06	275	−0.0221	0.5541	0.8165	0.0693	0.6694
4/25/06	276	−0.0247	0.1539	0.6564	0.0707	0.6683
5/2/06	277	−0.0235	0.3564	0.8795	0.0665	0.6671
5/9/06	278	−0.0186	0.8085	0.9092	0.0636	0.6652
5/16/06	279	−0.0197	0.5210	0.8617	0.0618	0.6634
5/23/06	280	−0.0200	0.5427	0.8246	0.0635	0.6621
5/30/06	281	−0.0223	0.3042	0.8321	0.0643	0.6617
6/6/06	282	−0.0198	0.3898	0.6811	0.0613	0.6611
6/13/06	283	−0.0229	0.1958	0.7353	0.0629	0.6612
6/20/06	284	−0.0235	0.4103	0.9313	0.0600	0.6613
6/27/06	285	−0.0242	0.4493	0.7293	0.0578	0.6613
7/4/06	286	−0.0243	0.1809	0.6872	0.0543	0.6612
7/11/06	287	−0.0222	0.3149	0.4469	0.0572	0.6614

(continued)

Table A.2 (continued)

Week start date	Week number	Assort.	Out. degree cent.	Inbound degree cent.	Between. cent.	Entropy
7/18/06	288	−0.0236	0.4570	0.6307	0.0529	0.6617
7/25/06	289	−0.0202	0.5070	0.4068	0.0583	0.6619
8/1/06	290	−0.0166	0.5788	0.7509	0.0622	0.6618
8/8/06	291	−0.0222	0.1934	0.6171	0.0594	0.6619
8/15/06	292	−0.0205	0.4004	0.3881	0.0550	0.6618
8/22/06	293	−0.0229	0.9042	0.9625	0.0509	0.6620
8/29/06	294	−0.0228	0.1529	0.7077	0.0524	0.6622
9/5/06	295	−0.0211	0.4093	0.7740	0.0531	0.6615
9/12/06	296	−0.0182	0.6849	0.9155	0.0514	0.6608
9/19/06	297	−0.0202	0.2595	0.6537	0.0511	0.6602
9/26/06	298	−0.0210	0.2861	0.7926	0.0487	0.6590
10/3/06	299	−0.0193	0.4783	0.6817	0.0475	0.6571
10/10/06	300	−0.0190	0.5285	0.7314	0.0483	0.6560
10/17/06	301	−0.0159	0.3505	0.7293	0.0467	0.6546
10/24/06	302	−0.0190	0.4697	0.9008	0.0462	0.6539
10/31/06	303	−0.0146	0.5130	0.5747	0.0467	0.6527
11/7/06	304	−0.0154	0.4131	0.7367	0.0463	0.6517
11/14/06	305	−0.0163	0.3489	0.8370	0.0456	0.6507
11/21/06	306	−0.0121	0.5504	0.5789	0.0404	0.6498
11/28/06	307	−0.0143	0.4472	0.4471	0.0426	0.6486
12/5/06	308	−0.0186	0.4540	0.6488	0.0478	0.6486
12/12/06	309	−0.0124	0.5342	0.6438	0.0404	0.6482
12/19/06	310	−0.0160	0.3509	0.6931	0.0367	0.6478
12/26/06	311	−0.0139	0.6247	0.6881	0.0376	0.6479
1/2/07	312	−0.0146	0.4139	0.4474	0.0411	0.6475
1/9/07	313	−0.0146	0.2381	0.6982	0.0424	0.6472
1/16/07	314	−0.0188	0.2736	0.6331	0.0446	0.6482
1/23/07	315	−0.0191	0.2960	0.6346	0.0445	0.6490
1/30/07	316	−0.0186	0.2412	0.6293	0.0466	0.6497
2/6/07	317	−0.0148	0.4343	0.4340	0.0375	0.6506
2/13/07	318	−0.0180	0.5684	0.8821	0.0421	0.6515
2/20/07	319	−0.0177	0.3066	0.6729	0.0430	0.6524
2/27/07	320	−0.0179	0.3941	0.7670	0.0408	0.6532
3/6/07	321	−0.0133	0.4898	0.7038	0.0401	0.6537
3/13/07	322	−0.0170	0.3222	0.6772	0.0371	0.6537
3/20/07	323	−0.0168	0.2472	0.4734	0.0368	0.6538
3/27/07	324	−0.0133	0.7752	0.7752	0.0347	0.6539
4/3/07	325	−0.0141	0.4211	0.4210	0.0346	0.6539
4/10/07	326	−0.0166	0.5249	0.8303	0.0370	0.6538
4/17/07	327	−0.0162	0.5011	0.5006	0.0342	0.6533
4/24/07	328	−0.0163	0.2751	0.7455	0.0342	0.6532

(continued)

Table A.2 (continued)

Week start date	Week number	Assort.	Out. degree cent.	Inbound degree cent.	Between. cent.	Entropy
5/1/07	329	−0.0163	0.5798	0.7985	0.0317	0.6529
5/8/07	330	−0.0144	0.3222	0.7859	0.0320	0.6525
5/15/07	331	−0.0167	0.4010	0.7966	0.0304	0.6522
5/22/07	332	−0.0149	0.2640	0.6235	0.0301	0.6521
5/29/07	333	−0.0148	0.2618	0.5497	0.0254	0.6521
6/5/07	334	−0.0142	0.3075	0.5903	0.0276	0.6520
6/12/07	335	−0.0162	0.2289	0.6822	0.0269	0.6519
6/19/07	336	−0.0161	0.2400	0.6230	0.0262	0.6518
6/26/07	337	−0.0171	0.5867	0.9479	0.0260	0.6519
7/3/07	338	−0.0189	0.9153	0.9702	0.0272	0.6522
7/10/07	339	−0.0183	0.5726	0.9654	0.0258	0.6525
7/17/07	340	−0.0177	0.3973	0.5196	0.0249	0.6528
7/24/07	341	−0.0192	0.3458	0.5789	0.0253	0.6529
7/31/07	342	−0.0180	0.6300	0.7614	0.0243	0.6532
8/7/07	343	−0.0168	0.3820	0.4939	0.0242	0.6533
8/14/07	344	−0.0179	0.3563	0.6748	0.0216	0.6534
8/21/07	345	−0.0167	0.2420	0.4442	0.0239	0.6534
8/28/07	346	−0.0183	0.5719	0.6873	0.0229	0.6533
9/4/07	347	−0.0160	0.6342	0.7491	0.0231	0.6530
9/11/07	348	−0.0095	0.5883	0.5623	0.0229	0.6524
9/18/07	349	−0.0161	0.3008	0.5885	0.0227	0.6518
9/25/07	350	−0.0155	0.5282	0.8212	0.0228	0.6513
10/2/07	351	−0.0054	0.8142	0.7830	0.0228	0.6507
10/9/07	352	−0.0145	0.2960	0.6018	0.0227	0.6502
10/16/07	353	−0.0166	0.2162	0.7227	0.0221	0.6498
10/23/07	354	−0.0117	0.7526	0.7731	0.0219	0.6494
10/30/07	355	−0.0111	0.8500	0.8500	0.0213	0.6492
11/6/07	356	−0.0155	0.5033	0.8974	0.0204	0.6486
11/13/07	357	−0.0130	0.6067	0.6073	0.0207	0.6483
11/20/07	358	−0.0127	0.4997	0.6783	0.0186	0.6480
11/27/07	359	−0.0163	0.2778	0.8493	0.0196	0.6475
12/4/07	360	−0.0134	0.2516	0.3821	0.0189	0.6471
12/11/07	361	−0.0143	0.3343	0.7122	0.0193	0.6467
12/18/07	362	−0.0130	0.3714	0.6151	0.0179	0.6465
12/25/07	363	−0.0135	0.3975	0.5818	0.0171	0.6464
1/1/08	364	−0.0139	0.2857	0.6825	0.0187	0.6459
1/8/08	365	−0.0145	0.3807	0.6325	0.0179	0.6451
1/15/08	366	−0.0093	0.5281	0.5266	0.0173	0.6444
1/22/08	367	−0.0118	0.5341	0.5164	0.0183	0.6438
1/29/08	368	−0.0090	0.4859	0.6154	0.0167	0.6429
2/5/08	369	−0.0102	0.4202	0.7099	0.0173	0.6421

(continued)

Table A.2 (continued)

Week start date	Week number	Assort.	Out. degree cent.	Inbound degree cent.	Between. cent.	Entropy
2/12/08	370	−0.0137	0.2737	0.8819	0.0179	0.6413
2/19/08	371	−0.0059	0.3432	0.2916	0.0176	0.6405
2/26/08	372	−0.0056	0.5076	0.2865	0.0173	0.6397
3/4/08	373	−0.0126	0.3213	0.8009	0.0170	0.6395
3/11/08	374	−0.0145	0.3914	0.7534	0.0169	0.6393
3/18/08	375	−0.0037	0.5329	0.5330	0.0165	0.6392
3/25/08	376	−0.0061	0.6000	0.5174	0.0169	0.6387
4/1/08	377	−0.0110	0.1402	0.3636	0.0169	0.6384
4/8/08	378	−0.0083	0.3741	0.6495	0.0171	0.6374
4/15/08	379	−0.0115	0.4704	0.4696	0.0172	0.6372
4/22/08	380	−0.0130	0.4373	0.9120	0.0155	0.6368
4/29/08	381	−0.0116	0.3492	0.6977	0.0157	0.6363
5/6/08	382	−0.0095	0.3486	0.5213	0.0152	0.6358
5/13/08	383	−0.0109	0.2493	0.6235	0.0151	0.6354
5/20/08	384	−0.0153	0.7224	0.9360	0.0150	0.6352
5/27/08	385	−0.0090	0.2856	0.4420	0.0136	0.6350
6/3/08	386	−0.0157	0.7258	0.8888	0.0148	0.6349
6/10/08	387	−0.0070	0.2798	0.2600	0.0145	0.6348
6/17/08	388	−0.0173	0.5289	0.7966	0.0146	0.6349
6/24/08	389	−0.0173	0.3020	0.6089	0.0132	0.6349
7/1/08	390	−0.0165	0.9512	0.9737	0.0142	0.6350
7/8/08	391	−0.0082	0.6504	0.6426	0.0137	0.6350
7/15/08	392	−0.0104	0.3924	0.4376	0.0135	0.6351
7/22/08	393	−0.0101	0.2869	0.3429	0.0132	0.6351
7/29/08	394	−0.0068	0.2965	0.2583	0.0124	0.6351
8/5/08	395	−0.0109	0.1932	0.3815	0.0134	0.6352
8/12/08	396	−0.0107	0.7237	0.9337	0.0130	0.6351
8/19/08	397	−0.0140	0.1934	0.5400	0.0122	0.6351
8/26/08	398	−0.0118	0.4585	0.6323	0.0131	0.6349
9/2/08	399	−0.0093	0.3460	0.4821	0.0130	0.6346
9/9/08	400	−0.0122	0.2975	0.6613	0.0130	0.6342
9/16/08	401	−0.0160	0.7125	0.9482	0.0128	0.6338
9/23/08	402	−0.0108	0.5635	0.8111	0.0133	0.6336
9/30/08	403	−0.0157	0.5540	0.9085	0.0125	0.6337
10/7/08	404	−0.0158	0.3403	0.5897	0.0128	0.6338
10/14/08	405	−0.0167	0.6384	0.8832	0.0137	0.6340
10/21/08	406	−0.0139	0.1486	0.5770	0.0133	0.6340
10/28/08	407	−0.0157	0.3262	0.5761	0.0131	0.6340
11/4/08	408	−0.0168	0.6306	0.9644	0.0131	0.6342
11/11/08	409	−0.0107	0.5604	0.5591	0.0141	0.6345
11/18/08	410	−0.0168	0.1078	0.6558	0.0137	0.6346
11/25/08	411	−0.0096	0.3481	0.4685	0.0118	0.6346

(continued)

Table A.2 (continued)

Week start date	Week number	Assort.	Out. degree cent.	Inbound degree cent.	Between. cent.	Entropy
12/2/08	412	−0.0154	0.7061	0.8091	0.0130	0.6345
12/9/08	413	−0.0144	0.1657	0.6311	0.0130	0.6343
12/16/08	414	−0.0175	0.3157	0.7545	0.0126	0.6345
12/23/08	415	−0.0161	0.6854	0.8586	0.0114	0.6346
12/30/08	416	−0.0129	0.2024	0.4163	0.0123	0.6348
1/6/09	417	−0.0093	0.5399	0.6244	0.0122	0.6345
1/13/09	418	−0.0113	0.5070	0.9240	0.0110	0.6340
1/20/09	419	−0.0134	0.5834	0.7551	0.0121	0.6336
1/27/09	420	−0.0129	0.5771	0.7043	0.0122	0.6331
2/3/09	421	−0.0125	0.2937	0.7657	0.0112	0.6326
2/10/09	422	−0.0168	0.5621	0.6474	0.0127	0.6325
2/17/09	423	−0.0151	0.5368	0.8128	0.0122	0.6323
2/24/09	424	−0.0067	0.5356	0.5348	0.0114	0.6318
3/3/09	425	−0.0132	0.4317	0.7655	0.0110	0.6315
3/10/09	426	−0.0151	0.5238	0.5375	0.0123	0.6315
3/17/09	427	−0.0105	0.3631	0.5621	0.0114	0.6314
3/24/09	428	−0.0094	0.4562	0.6094	0.0113	0.6314
3/31/09	429	−0.0103	0.4243	0.6504	0.0111	0.6315
4/7/09	430	−0.0098	0.4565	0.5983	0.0111	0.6315
4/14/09	431	−0.0133	0.3330	0.6270	0.0113	0.6315
4/21/09	432	−0.0110	0.2535	0.4828	0.0116	0.6316
4/28/09	433	−0.0086	0.6923	0.7246	0.0111	0.6316
5/5/09	434	−0.0127	0.3337	0.6309	0.0115	0.6317
5/12/09	435	−0.0112	0.3488	0.6062	0.0109	0.6317
5/19/09	436	−0.0118	0.3025	0.5834	0.0101	0.6317
5/26/09	437	−0.0126	0.1778	0.4651	0.0113	0.6318
6/2/09	438	−0.0101	0.4887	0.6906	0.0108	0.6318
6/9/09	439	−0.0089	0.4450	0.5176	0.0110	0.6319
6/16/09	440	−0.0164	0.2401	0.7721	0.0115	0.6321
6/23/09	441	−0.0128	0.1961	0.4107	0.0105	0.6323
6/30/09	442	−0.0129	0.2654	0.4902	0.0105	0.6324
7/7/09	443	−0.0114	0.4749	0.7882	0.0106	0.6325
7/14/09	444	−0.0104	0.3377	0.5750	0.0100	0.6326
7/21/09	445	−0.0084	0.5087	0.5082	0.0104	0.6327
7/28/09	446	−0.0092	0.4209	0.5091	0.0103	0.6327
8/4/09	447	−0.0109	0.4188	0.6329	0.0102	0.6328
8/11/09	448	−0.0092	0.4117	0.5305	0.0095	0.6329
8/18/09	449	−0.0118	0.3443	0.5876	0.0099	0.6330
8/25/09	450	−0.0120	0.3249	0.6453	0.0102	0.6331
9/1/09	451	−0.0091	0.3288	0.4240	0.0096	0.6332
9/8/09	452	−0.0157	0.0574	0.5789	0.0106	0.6334
9/15/09	453	−0.0109	0.4223	0.6520	0.0098	0.6334

(continued)

Table A.2 (continued)

Week start date	Week number	Assort.	Out. degree cent.	Inbound degree cent.	Between. cent.	Entropy
9/22/09	454	−0.0075	0.4526	0.5042	0.0094	0.6334
9/29/09	455	−0.0137	0.1758	0.6375	0.0103	0.6335
10/6/09	456	−0.0122	0.2508	0.6751	0.0099	0.6335
10/13/09	457	−0.0107	0.4258	0.7217	0.0097	0.6335
10/20/09	458	−0.0107	0.3557	0.6073	0.0101	0.6335
10/27/09	459	−0.0113	0.3401	0.6653	0.0100	0.6335
11/3/09	460	−0.0098	0.4006	0.6515	0.0093	0.6335
11/10/09	461	−0.0136	0.2641	0.6513	0.0089	0.6335
11/17/09	462	−0.0151	0.1319	0.6469	0.0105	0.6337
11/24/09	463	−0.0163	0.3832	0.5884	0.0097	0.6339
12/1/09	464	−0.0145	0.1439	0.5050	0.0099	0.6340
12/8/09	465	−0.0139	0.1408	0.6310	0.0094	0.6341
12/15/09	466	−0.0145	0.5161	0.9198	0.0090	0.6341
12/22/09	467	−0.0168	0.1220	0.6518	0.0080	0.6343
12/29/09	468	−0.0126	0.3479	0.5958	0.0090	0.6344
1/5/10	469	−0.0120	0.2776	0.5870	0.0089	0.6344
1/12/10	470	−0.0122	0.2775	0.6303	0.0093	0.6345
1/19/10	471	−0.0117	0.2531	0.5499	0.0100	0.6345
1/26/10	472	−0.0124	0.2356	0.6002	0.0098	0.6346
2/2/10	473	−0.0054	0.9423	0.9105	0.0088	0.6345
2/9/10	474	−0.0119	0.2806	0.5654	0.0093	0.6345
2/16/10	475	−0.0156	0.1142	0.6485	0.0100	0.6345
2/23/10	476	−0.0141	0.1306	0.7804	0.0092	0.6345
3/2/10	477	−0.0121	0.1844	0.6229	0.0089	0.6345
3/9/10	478	−0.0140	0.1927	0.7157	0.0086	0.6345
3/16/10	479	−0.0145	0.4126	0.5996	0.0085	0.6345
3/23/10	480	−0.0109	0.2899	0.5295	0.0078	0.6345
3/30/10	481	−0.0105	0.4637	0.4628	0.0076	0.6346
4/6/10	482	−0.0147	0.1995	0.7363	0.0085	0.6346
4/13/10	483	−0.0143	0.1806	0.7194	0.0085	0.6346
4/20/10	484	−0.0122	0.2327	0.7580	0.0086	0.6346
4/27/10	485	−0.0096	0.6680	0.8899	0.0082	0.6346
5/4/10	486	−0.0111	0.4092	0.8156	0.0079	0.6347
5/11/10	487	−0.0116	0.3202	0.6344	0.0090	0.6347
5/18/10	488	−0.0111	0.4756	0.7533	0.0078	0.6347
5/25/10	489	−0.0142	0.3630	0.7850	0.0078	0.6348
6/1/10	490	−0.0144	0.3325	0.7857	0.0067	0.6348
6/8/10	491	−0.0150	0.1611	0.6029	0.0075	0.6349
6/15/10	492	−0.0127	0.2810	0.8708	0.0076	0.6349
6/22/10	493	−0.0104	0.2803	0.6184	0.0080	0.6349
6/29/10	494	−0.0116	0.2304	0.7240	0.0071	0.6349
7/6/10	495	Not est.	Not est.	Not est.	Not est.	0.6349

Table A.3 Entropy and stickiness of logged-in editors

Week number	Logged-in editors' entropy	Registered elite stickiness (Lag 1)
4	0.7940	66.67
9	0.6898	40.00
14	0.7618	20.00
19	0.7682	37.50
24	0.7666	22.22
29	0.7551	20.00
34	0.7722	100.00
39	0.7897	37.04
44	0.7948	61.90
49	0.7788	54.55
54	0.7679	56.41
59	0.7485	15.00
64	0.7441	66.67
69	0.7364	57.14
74	0.7334	64.10
79	0.7284	60.47
84	0.7243	31.71
89	0.7012	5.88
94	0.4925	100.00
99	0.4978	70.00
104	0.5162	73.68
109	0.5272	67.69
114	0.5414	58.67
119	0.5526	78.08
124	0.5649	9.35
129	0.5691	94.44
134	0.5754	71.59
139	0.5871	57.43
144	0.5948	54.67
149	0.6034	56.72
154	0.6128	50.39
159	0.6209	86.17
164	0.6298	34.78
169	0.6416	71.88
174	0.6506	31.48
179	0.6571	73.97
184	0.6644	56.11
189	0.6719	33.95
194	0.6769	56.40
199	0.6827	66.23
204	0.6876	59.41
209	0.6968	56.71

(continued)

Table A.3 (continued)

Week number	Logged-in editors' entropy	Registered elite stickiness (Lag 1)
214	0.7050	51.04
219	0.7103	32.29
224	0.7166	54.73
229	0.7175	52.17
234	0.7164	50.70
239	0.7155	34.73
244	0.7130	53.74
249	0.7082	58.43
254	0.7013	44.54
259	0.6933	29.80
264	0.6831	52.52
269	0.6769	38.41
274	0.6709	37.73
279	0.6634	42.44
284	0.6613	46.63
289	0.6619	35.14
294	0.6622	29.03
299	0.6571	42.93
304	0.6517	44.75
309	0.6482	48.36
314	0.6482	54.00
319	0.6524	26.33
324	0.6539	32.40
329	0.6529	35.83
334	0.6520	46.81
339	0.6525	35.97
344	0.6534	29.81
349	0.6518	50.00
354	0.6494	37.28
359	0.6475	32.53
364	0.6459	37.50
369	0.6421	44.55
374	0.6393	39.53
379	0.6372	42.25
384	0.6352	37.12
389	0.6349	38.41
394	0.6351	37.28
399	0.6346	54.55
404	0.6338	52.67
409	0.6345	34.08
414	0.6345	28.54
419	0.6336	43.69

(continued)

Table A.3 (continued)

Week number	Logged-in editors' entropy	Registered elite stickiness (Lag 1)
424	0.6318	41.70
429	0.6315	42.62
434	0.6317	53.58
439	0.6319	44.99
444	0.6326	41.05
449	0.6330	53.16
454	0.6334	30.64
459	0.6335	51.05
464	0.6340	43.30
469	0.6344	45.69
474	0.6345	41.13
479	0.6345	37.50
484	0.6346	41.51
489	0.6348	39.12

Organizational Structure Measures

Part of our research deals with the gradual trends and sudden phase shifts that the Wikipedia community underwent. In order to properly address this question, the five well-established measures of organizational structure and behavior described above were assessed in 1-week windows for a total of 491 measurements.

First, structural changes in collaborative attractiveness, as they occurred in the Wikipedia community, were assessed using inbound degree centralization. Freeman (1979) developed this metric to indicate the extent to which some actors in a given network are more attractive than others. Broadly speaking, the more edges that point toward a given actor, and the stronger those edges are, the more effort others are exerting to form a relationship with that individual. The more that a few key individuals stand out as being substantially more "central" than others, the more centralized the organization is around those individuals.

In the Wikipedia context, it is quite possible that the revisions made by a few central editors are, for whatever reason, especially attractive to others, in that their revisions to particular articles prompt other editors to further revise and develop the content in question. It would be as though the other community members were clamoring around the few "popular" editors in question, following them and further editing their content across the knowledge production system. The small locus of attractive editors would thus hold a great deal of influence in dictating where the community at large focused its efforts.

Simply put, inbound degree centralization is designed to measure the extent of stratification between the "popular" group and the rest of the masses. Higher values of inbound degree centralization indicate that a small set of editors are far more attractive than the rest, with fellow editors much more inclined to engage with the content added by the attractive few than that contributed by others. In such a community, the power is localized around those individuals toward whom the community at large is drawn.

Likewise, changes in the structure of collaborative extroversion were represented by outbound degree centralization. Using a similar approach as that of inbound degree centralization, outbound degree centralization examines how active an individual is in forming and strengthening connections with others in the network by summing the weight of all edges leading away from the actor in question toward collaborative peers. Those individuals who form strong relationships with many others can influence the behavior of a wide proportion of the population accordingly.

The extent to which some individuals are more active than others in expanding their span of influence is measured using outbound degree centralization. As outbound degree centralization increases, the locus of power becomes more concentrated within a few individuals who are far more active than others in forging connections with their peers and spreading their influence throughout the organization.

Next, changes in communication flow through the community were operationalized as shifts in betweenness centralization. As noted by Freeman (1979), among others, an individual lying on the shortest path between any two others in a network holds a great deal of influence over the relationship between those two individuals, as ideas and knowledge that would move from one to the other are likely to pass through that central individual. In other words, the person along the shortest path between two other individuals acts as a gatekeeper between them. In the Wikipedia network, this manifests itself in control over the flow of information or knowledge, which is the most prized resource within knowledge construction communities.

Betweenness centralization assesses the localization of such power, specifically by examining the disparity between the most central community members and everyone else in the network. Higher betweenness centralization indicates that the locus of power is being leeched away from the network at large as a few key individuals take command over the flow of information throughout the entire organization. It also indicates a higher potential velocity of communication and collaboration flows, with a small group of elites acting as hubs responsible for transmitting information throughout the organization.

Trends in partner choice, on the other hand, were evaluated using assortativity. This measure highlights the choices that fellow community members make when forming connections with others who are similar or dissimilar to themselves (Newman 2002). In other words, when the general trend is for organizational members to form coediting connections with others who share a similar level of prominence as themselves—whether they are similarly well connected or similarly poorly connected in the community—assortativity increases. As this trend declines, assortativity itself declines. Should the trend reverse itself, such that community members seek out coediting connections with others who hold dissimilar levels of prominence to their own more than they pursue those of a similar level, then the assortativity measure becomes negative, indicating a disassortative tendency throughout the organization.

On Wikipedia, past research (Britt 2011) indicated that the community behaved in a disassortative manner during the early weeks of its development, but that this trend progressively declined. However, as noted above, this previous study aggregated the network of contributions over time, so the disassortativity finding suggesting that community members sought out dissimilar partners was due, at least in part, to the positions they had established over the prior weeks, months, and years. It is quite possible that the network actually became assortative in later weeks, but that this change was obscured by the structure of the past. As previously noted, the research presented in this volume avoids such residual effects in order to make any such changes more visible.

Finally, social entropy served as a measurement of the extent of order within the community. In contrast with the various forms of centralization and assortativity, social entropy was measured using the distribution of contributions as opposed to the network of connections between editors (see Matei et al. 2015; Matei et al. 2010a). The significance of each revision made by a given editor, as quantified by de Alfaro et al.'s (2010) delta measure, was used to assess how much that editor contributed to the community, with the delta values of all revisions the editor made in a given week summed to determine his or her overall contribution to the project in that week (Matei et al. 2015). The more equal that editor contributions across the community were during a given week, the higher that the community's social entropy climbed toward a maximum value of 1, signifying perfect evenness. Likewise, unequal levels of contributions would cause the social entropy calculation to drop, with the minimum of zero indicating that only one community member made any contributions whatsoever during the week in question (Matei et al. 2015).

Existing Breakpoint Identification Approaches

The present study uses a new approach to identify the locations of breakpoints. As such, before delving into the details of the present analysis, it is worth explaining why other, established techniques would have been inappropriate for this study.

First, it would be reasonable to expect the time series data in the present study to exhibit some autocorrelative properties, which might suggest the use of an ARIMA model, a spectral analysis, or a similar statistical approach. However, the key components of an ARIMA model have the potential to mask phase shifts, effectively negating the very trends we hope to observe. For instance, the entire point of differencing data is to remove linear and higher-order terms, which can make it extremely difficult to detect the locations of otherwise clear changes in the trend over time.

Consider, for instance, the fictional data set in Fig. A.1, which was generated from two distinct linear trends (given by the red and blue lines) along with normally distributed random noise. In Fig. A.1, the trend appears to be fairly constant for the first 100 data points. Around that 100th data point, however, its value plummets, and then the fictional measure appears to increase in a roughly linear fashion for the remainder of the data set. This radical change is easy to observe, even with the naked eye.

Figure A.2, however, shows the result of differencing the example data. From this plot, there does not appear to be any change at all, despite how obvious the

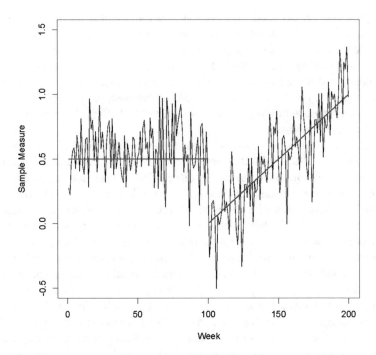

Fig. A.1 Fictional data generated from two normally distributed linear trends with the trends plotted in *red* and *blue*

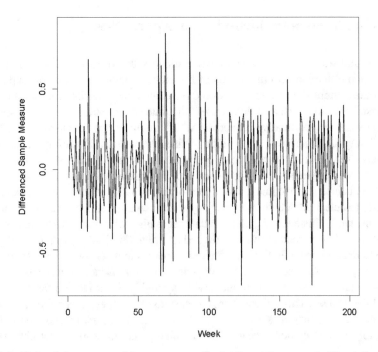

Fig. A.2 Fictional data generated from two normally distributed linear trends (Fig. A.1), differenced once

transition was in Fig. A.1. Now that the data has been differenced, it would be virtually impossible to detect the 100th data point as a key change moment, regardless of the statistical approach employed.

It is clear that while differencing would be necessary to fit autoregressive or moving average terms to the present data, doing so can also mask moments of flux and substantially inhibit any subsequent effort to detect changes. Furthermore, ARMA terms alone can obscure otherwise distinct breakpoints. Moving averages, after all, are frequently used for smoothing jagged plots, which in this case may be the most crucial part of the data. Even when a breakpoint is still evident after applying a moving average term, its location may be left in doubt, as the analysis of any given data point will necessarily include other data points as well, blending them together and making it difficult to determine at which data point a change occurred. Autoregressive terms can have similar effects. At best, ARIMA terms could mask the exact location of a given breakpoint, and at worst, they could hide the breakpoint entirely.

A few other techniques exist to handle problems related to change points in data, but each of these suffers from a critical limitation. Segmented regression analysis, which is generally used to detect repeated-samples treatment effects in quasi-experiments, presupposes that the location of the change is known and only tests the significance of the change at that location, whether in terms of the intercept, the slope, or a higher-order term. In other words, segmented regression merely assesses the differences between two or more known segments. In the present study, no experimental change was deliberately applied by any researcher—rather, any changes were observed as they naturally occurred without intervention—so any change points that may have occurred were not known in advance. Therefore, since we could not assume any particular moments of change, our goal was instead to identify the significant breakpoints within the data set whose locations and number were unknown.

Several iterative methods have been developed to find the most appropriate breakpoint locations given the number of breakpoints for which to search. The SAS statistical package, for instance, uses steepest-descent and gradient descent approaches as well as the Newton, Gauss-Newton, and Marquardt methods of estimating breakpoints. However, these iterative methods can only be used on continuous segments. In other words, all of these approaches assume that data segments are linked together and only aim to detect a "bend" in a continuous line, such as that shown in Fig. A.3. As such, they are unable to identify "jumps" in the data caused by an instantaneous change in the intercept when the higher-order terms remain constant between the two segments (Fig. A.4), a scenario that was considered to be likely in the present study (see Britt 2013).

As a final alternative, a team of statisticians led by Achim Zeileis (Zeileis et al. 2003, 2010) devised the "strucchange" algorithm to determine the number and locations of N unknown breakpoints by minimizing the Bayesian information criterion (BIC). However, initial testing indicated that this approach tended to dramatically overfit data when its variance was high in comparison with evolutionary movement due to linear and higher-order terms. This issue was so severe that in almost every

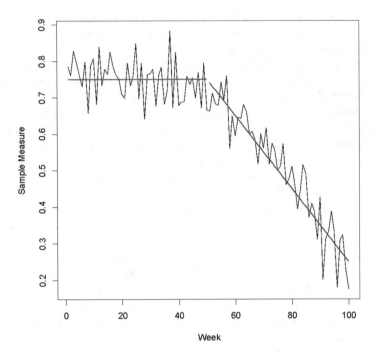

Fig. A.3 Fictional data generated from two normally distributed linear trends with a continuous breakpoint, with the trends plotted in *red* and *blue*

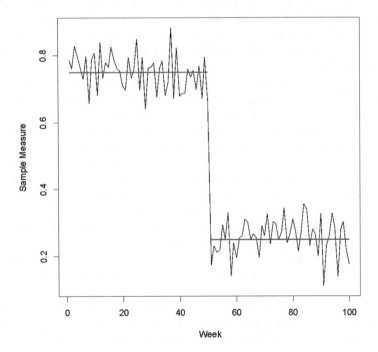

Fig. A.4 Fictional data generated from two normally distributed linear trends with a discontinuous breakpoint, with the trends plotted in *red* and *blue*

trial there were as many regression terms as data points assigned to some segments. The problem was further exacerbated when the variance was heteroscedastic, as Zeileis's algorithm flooded the more volatile areas of the data set with numerous breakpoints, perfectly fitting random noise and generating comparatively mediocre fits for the more significant, meaningful shifts that occurred during periods of lesser variance. Britt (2011) previously found that assortativity on Wikipedia followed an especially heteroscedastic trend, and other factors addressed in this volume were later revealed to have similar variance structures, so Zeileis's approach would not have been appropriate for this study.

Stepwise Segmented Regression Analysis

Algorithm Description

With the inadequacy of the above approaches in mind, this analysis used an alternative, more general approach based on the principles of stepwise regression model selection. This approach was inspired by Crawley's (2007, 425–430) method of conducting a piecewise regression using indicator functions (see also Lemoine 2012). However, his approach only incorporated intercepts and linear terms, neglecting the potential for curvilinear trends, and it focused only on detecting significant trends for each segment rather than assessing the significance of the breakpoints that divided them. The present study synthesizes and builds upon Crawley's initial ideas in order to detect revolutionary changes, not just the significant evolutionary trends between them, and to handle multiple breakpoints and higher-order regression terms.

Using the R statistical software package, a separate regression model was fitted to each of the five dependent variables measured in this study: inbound degree centralization, outbound degree centralization, betweenness centralization, assortativity, and entropy. For each of these models, an intercept term, the slope, and a quadratic term were included in the model, and these base terms, much like the first level in a hierarchical regression, were never removed. Afterward, interaction terms incorporating indicator functions were added and removed from the model based on the $\alpha = .15$ standard, a common threshold for stepwise model building. At each stage of the process, terms were added to the model if they were statistically significant at the $\alpha = .15$ level, then insignificant terms were removed if they no longer met the $\alpha = .15$ threshold after the addition of other terms.

Each indicator function corresponded to all data points prior to a specified week in the data set. If a given observation of the measure in question occurred after that specified week, the indicator function took on a value of 1. Otherwise, the indicator function was 0. For instance, if an interaction term corresponding to week 100 was added to the model for inbound degree centralization, then the $Intercept^*I(Week \geq 100)$ term would equal 1 for all observations of inbound degree centralization from week 100 onward. For weeks 4–99, the $I(Week \geq 100)$ indicator function was 0, so the $Intercept^*I(Week \geq 100)$ term estimated the intercept from week 100 onward, in addition to the global $Intercept$ term applied to the entire data set. As such, the interaction term effectively served to estimate the change in the intercept at each breakpoint. The same approach applied for the linear and quadratic terms in each regression model. Each indicator function was treated as its own variable, multiplied by the week number, and multiplied by the squared week number, such that the changes in the intercept, slope, and quadratic terms at each breakpoint were each estimated by these various interaction terms.

A given week was deemed a significant breakpoint whenever its indicator function, its indicator function times the week number, or its indicator function times the week number squared were all added to the model and at least one of the three terms was statistically significant. In other words, if adding a breakpoint to a data segment would split it into two segments with significantly different intercepts, slopes, or quadratic terms at the $\alpha = .15$ level, all three were added to the regression model accordingly. Likewise, only when all three terms were insignificant was the breakpoint removed from the model.

Terms were grouped in this way in order to prevent needlessly overfitting the number of breakpoints due to spurious statistics. Otherwise, one could envision, for instance, a slope change being added to the model in the 100th week, and a quadratic change later being added at week 101—not because both weeks served as independent breakpoints modeling distinct changes on Wikipedia but merely because fitting the second change to week 101 would happen to explain more variance than adding a second indicator function for week 100 to the model. This, notably, would result in the higher-order terms between the two breakpoints being inestimable.

It is also worth noting that two breakpoints corresponding to back-to-back weeks could nonetheless be added to the model on the sole basis of their respective intercepts being significantly different in their own right. This would also result in the higher-order terms corresponding to short line segments being inestimable, but this would nonetheless cause no problems with the core task of identifying the locations of statistically significant breakpoints, so it was otherwise not a problem for model fitting.

Step-by-Step Walkthrough

As an example of this process, the model-building process for Measure A begins with an intercept, slope, and quadratic term already in the model:

$$Y = Intercept + Week + Week^2 \tag{A.3}$$

Then, the model-building process considers adding a new intercept, slope, and quadratic term to the model, each of which would be multiplied by an intercept term:

$$Y = Intercept + Week + Week^2 + Intercept^*I\left(Week \geq X\right)$$
$$+Week^*I\left(Week \geq X\right) + Week^{2*}I\left(Week \geq X\right) \tag{A.4}$$

The process iteratively builds the regression model with X equal to every week number in the data set (with the obvious exception of the last week, since no observations could have a higher week number than the last week in the data set). Then it selects the value of X that would yield the regression model with the smallest error sum of squares, and if $Intercept^*I(Week \geq X)$, $Week^*I(Week \geq X)$, or $Week^{2*}I(Week \geq X)$ is statistically significant at the $\alpha = .15$ level, all three terms are added to the model. For the sake of the example, suppose that the value of X that would yield the smallest error sum of squares is $X = 300$. Then the process would consider the model

$$Y = Intercept + Week + Week^2 + Intercept^*I\left(Week \geq 300\right)$$
$$+Week^*I\left(Week \geq 300\right) + Week^{2*}I\left(Week \geq 300\right) \tag{A.5}$$

and if $Intercept^*I(Week \geq 300)$, $Week^*I(Week \geq 300)$, or $Week^{2*}I(Week \geq 300)$ was statistically significant at the $\alpha = .15$ level, this model would be accepted, and the process would move to the next step. For instance, if $Week^*I(Week \geq 300)$ was statistically significant, that would mean that the slope of the segment ending in week 299 was significantly different than the slope of the segment from week 300 onward.

After the three terms corresponding to a breakpoint at a given week were added to the model, any insignificant breakpoints were removed from the model. For instance, consider the above model a few steps later in the model-building process:

$$Y = Intercept + Week + Week^2 + Intercept^*I\left(Week \geq 190\right) + Week^*I\left(Week \geq 190\right)$$
$$+Week^{2*}I\left(Week \geq 190\right) + Intercept^*I\left(Week \geq 200\right) + Week^*I\left(Week \geq 200\right)$$
$$+Week^{2*}I\left(Week \geq 200\right) + Intercept^*I\left(Week \geq 300\right) + Week^*I\left(Week \geq 300\right)$$
$$+Week^{2*}I\left(Week \geq 300\right) + Intercept^*I\left(Week \geq 310\right) + Week^*I\left(Week \geq 310\right)$$
$$+Week^{2*}I\left(Week \geq 310\right) \tag{A.6}$$

After the latest set of intercept, slope, and quadratic terms are added to the model, all terms are again checked for significance. If all three terms corresponding to a given week have become statistically insignificant, those three terms are removed from the model. Suppose that in this case $Intercept^*I(Week \geq 200)$, $Week^*I(Week \geq 200)$, and $Week^{2*}I(Week \geq 200)$ are all statistically insignificant terms at the $\alpha = .15$ level, and $Intercept^*I(Week \geq 300)$, $Week^*I(Week \geq 300)$, and $Week^{2*}I(Week \geq 300)$ are also statistically insignificant. In that case, the model becomes

$$Y = Intercept + Week + Week^2 + Intercept^*I\left(Week \geq 190\right) + Week^*I\left(Week \geq 190\right)$$
$$+Week^{2*}I\left(Week \geq 190\right) + Intercept^*I\left(Week \geq 310\right) + Week^*I\left(Week \geq 310\right)$$
$$+Week^{2*}I\left(Week \geq 310\right) \tag{A.7}$$

and the process will next determine whether the terms for another breakpoint can be added to the model.

In this project, the process rotated back and forth between these two steps, adding the terms for a significant breakpoint, removing the terms for any insignificant breakpoints, and then adding the terms for another significant breakpoint, until no further terms could be added or removed from the model.

It should also be noted that the independent variables were certainly not independent from one another, as adding or removing any indicator function inevitably changed the number of data points estimated by the term which followed, which had the potential to profoundly affect the existing estimation. This multicollinearity was inescapable but in this case was not cause for alarm. After all, the goal was to find breakpoints indicating significant differences between adjacent data segments. If the potential effect of an indicator term was masked due to multicollinearity, that only meant that the term in question did not highlight a substantial change from the trends already in the model and therefore did not deserve to be added as a breakpoint in its own right. In much the same way, typical regression diagnostics for phenomena like outliers and non-normality would have been meaningless, as the entire point of this analysis was to detect significant deviations from the existing model.

Once the final model was determined, the respective significance of each term was assessed at the standard $\alpha = .05$ level. Because of the large number of factors under consideration, a Holm-Bonferroni correction was applied to the model selection and significance testing thresholds in order to control the experimentwise Type I error rate. With that in mind, in addition to the three terms included from the beginning, a total of 487 breakpoints had the potential to be added to each regression model. Each of these breakpoints, in turn, was comprised of three terms that could be significant, the intercept, the slope, and the quadratic term, and if any one of those terms was significant, all three were added to the regression model. As such, the Holm-Bonferroni correction was applied based on the presence of 1,464 regression terms being tested.

Results

The breakpoint detection procedure in our study used stepwise segmented regression models for each of the five respective organizational measures. These models are given in Tables A.4, A.5, A.6, A.7, and A.8, and they correspond to the time series plots in Figs. A.5, A.6, A.7, A.8, and A.9. Significant factors at the $\alpha = .05$ level, with the Holm-Bonferroni correction, are labeled with an asterisk. Figure A.10 features all five measures on a single plot, without the fitted values from their respective regression models.

For each regression model in Tables A.4, A.5, A.6, A.7, and A.8, each segment is defined by the sum of all applicable indicator functions. For instance, in the model of outbound degree centralization (Table A.5, Fig. A.6), the segment from weeks 4 to 141 is defined by the equation

$$Outbound\,Degree\,Centralization$$
$$= 7.674E^{-1} - 5.608E^{-3*}Week + 1.720E^{-5*}Week^2, \qquad (A.8)$$

the segment from weeks 142 to 206 is given by

Table A.4 Regression model for inbound degree centralization

	Estimate	Std. error	T-value	P-value	
Intercept	$6.112E^{-1}$	$2.686E^{-2}$	22.753	$<2E^{-16}$	*
Week	$3.172E^{-4}$	$2.489E^{-4}$	1.274	0.203	
Week^2	$-4.798E^{-7}$	$4.844E^{-7}$	-0.990	0.322	

*Denotes statistically significant variables ($\alpha = .05$)

Table A.5 Regression model for outbound degree centralization

	Estimate	Std. error	T-value	P-value	
Intercept	$7.674E^{-1}$	$5.671E^{-2}$	13.533	$<2E^{-16}$	*
Intercept*I(Week $\geq$ 142)	$1.254E^1$	$2.381E^0$	5.268	$2.08E^{-7}$	*
Intercept*I(Week $\geq$ 207)	$-1.380E^1$	$2.391E^0$	-5.771	$1.41E^{-8}$	*
Week	$-5.608E^{-3}$	$1.799E^{-3}$	-3.117	0.00194	*
Week*I(Week $\geq$ 142)	$-1.473E^{-1}$	$2.764E^{-2}$	-5.329	$1.52E^{-7}$	*
Week*I(Week $\geq$ 207)	$1.583E^{-1}$	$2.762E^{-2}$	5.734	$1.74E^{-8}$	*
Week^2	$1.720E^{-5}$	$1.205E^{-5}$	1.428	0.15397	
Week^2*I(Week $\geq$ 142)	$4.290E^{-4}$	$8.008E^{-5}$	5.357	$1.31E^{-7}$	*
Week^2*I(Week $\geq$ 207)	$-4.541E^{-4}$	$7.919E^{-5}$	-5.734	$1.74E^{-8}$	*

*Denotes statistically significant variables ($\alpha = .05$)

Table A.6 Regression model for betweenness centralization

	Estimate	Std. error	T-value	P-value	
Intercept	$2.029E^{-1}$	$1.799E^{-2}$	11.274	$<2E^{-16}$	*
Intercept*I(Week $\geq$ 54)	$1.297E^{-1}$	$1.939E^{-2}$	6.689	$6.20E^{-11}$	*
Week	$-5.602E^{-3}$	$1.435E^{-3}$	-3.904	$1.08E^{-4}$	*
Week*I(Week $\geq$ 54)	$4.201E^{-3}$	$1.436E^{-3}$	2.925	$3.61E^{-3}$	
Week^2	$2.062E^{-4}$	$2.457E^{-5}$	8.390	$5.37E^{-16}$	*
Week^2*I(Week $\geq$ 54)	$-2.047E^{-4}$	$2.457E^{-5}$	-8.328	$8.49E^{-16}$	*

*Denotes statistically significant variables ($\alpha = .05$)

Table A.7 Regression model for assortativity

	Estimate	Std. error	T-value	P-value	
Intercept	$-2.802E^{-1}$	$1.697E^{-2}$	-16.513	$<2E^{-16}$	*
Week	$1.426E^{-3}$	$1.569E^{-4}$	9.089	$< 2E^{-16}$	*
Week^2	$-1.846E^{-6}$	$3.048E^{-7}$	-6.057	$2.78E^{-9}$	*

*Denotes statistically significant variables ($\alpha = .05$)

$$Outbound\ Degree\ Centralization = 7.674E^{-1} + 1.254E^{+1}$$
$$+ \left(-5.608E^{-3} - 1.473E^{-1}\right)^{*} Week + \left(1.720E^{-5} + 4.290E^{-4}\right)^{*} Week^{2}, \quad \text{(A.9)}$$

and the segment from weeks 207 to 491 follows

Table A.8 Regression model for entropy

	Estimate	Std. error	T-value	P-value	
Intercept	$3.408E^{-1}$	$1.034E^{-1}$	3.297	0.00105	
Intercept*I(Week $\geq$ 7)	$-2.377E^{0}$	$2.872E^{-1}$	-8.277	$1.34E^{-15}$	*
Intercept*I(Week $\geq$ 10)	$2.825E^{0}$	$2.679E^{-1}$	10.542	$<2E^{-16}$	*
Intercept*I(Week $\geq$ 42)	$1.559E^{-1}$	$1.227E^{-2}$	12.711	$<2E^{-16}$	*
Intercept*I(Week $\geq$ 92)	$-6.323E^{-1}$	$1.358E^{-2}$	-46.574	$<2E^{-16}$	*
Intercept*I(Week $\geq$ 93)	$-8.551E^{-2}$	$3.586E^{-3}$	-23.843	$<2E^{-16}$	*
Intercept*I(Week $\geq$ 204)	$-1.623E^{0}$	$1.652E^{-1}$	-9.822	$<2E^{-16}$	*
Intercept*I(Week $\geq$ 250)	$3.325E^{0}$	$1.753E^{-1}$	18.974	$<2E^{-16}$	*
Intercept*I(Week $\geq$ 335)	$-8.865E^{-1}$	$6.376E^{-2}$	-13.905	$<2E^{-16}$	*
Week	$2.167E^{-1}$	$4.226E^{-2}$	5.127	$4.31E^{-7}$	*
Week*I(Week $\geq$ 7)	$4.734E^{-1}$	$7.967E^{-2}$	5.942	$5.52E^{-9}$	*
Week*I(Week $\geq$ 10)	$-6.932E^{-1}$	$6.754E^{-2}$	-10.263	$<2E^{-16}$	*
Week*I(Week $\geq$ 42)	$-1.301E^{-3}$	$5.413E^{-4}$	-2.403	0.01664	
Week*I(Week $\geq$ 92)	$7.719E^{-3}$	$3.654E^{-4}$	21.128	$<2E^{-16}$	*
Week*I(Week $\geq$ 204)	$1.489E^{-2}$	$1.465E^{-3}$	10.167	$<2E^{-16}$	*
Week*I(Week $\geq$ 250)	$-2.631E^{-2}$	$1.516E^{-3}$	-17.354	$<2E^{-16}$	*
Week*I(Week $\geq$ 335)	$6.247E^{-3}$	$4.225E^{-4}$	14.784	$<2E^{-16}$	*
Week^{2}	$-2.585E^{-2}$	$4.219E^{-3}$	-6.127	$1.91E^{-9}$	*
Week^{2}*I(Week $\geq$ 7)	$-1.718E^{-2}$	$5.966E^{-3}$	-10.263	0.00417	
Week^{2}*I(Week $\geq$ 10)	$4.310E^{-2}$	$4.219E^{-3}$	10.217	$<2E^{-16}$	*
Week^{2}*I(Week $\geq$ 42)	$-5.709E^{-5}$	$8.415E^{-6}$	-6.784	$3.55E^{-11}$	*
Week^{2}*I(Week $\geq$ 92)	$-2.550E^{-5}$	$2.641E^{-6}$	-9.656	$<2E^{-16}$	*
Week^{2}*I(Week $\geq$ 204)	$-3.408E^{-5}$	$3.243E^{-6}$	-10.506	$<2E^{-16}$	*
Week^{2}*I(Week $\geq$ 250)	$5.204E^{-5}$	$3.298E^{-6}$	15.780	$<2E^{-16}$	*
Week^{2}*I(Week $\geq$ 335)	$-1.075E^{-5}$	$7.085E^{-7}$	-15.172	$<2E^{-16}$	*

*Denotes statistically significant variables ($\alpha = .05$)

$$Outbound\ Degree\ Centralization = 7.674E^{-1} + 1.254E^{+1}$$
$$-1.380E^{+1} + \left(-5.608E^{-3} - 1.473E^{-1} + 1.583E^{-1}\right)^{*}$$
$$Week + \left(1.720E^{-5} + 4.290E^{-4} - 4.541E^{-4}\right)^{*} Week^{2}$$

(A.10)

with E used to represent scientific notation as appropriate. In other words, each coefficient indicated the magnitude of the shift at each breakpoint. The estimated values of the intercept, slope, and quadratic term themselves may be determined by taking each sequential shift into account.

For convenience, breakpoints are named by the week number that begins the segment following the breakpoint. For instance, the outbound degree centralization model defined two breakpoints positioned between weeks 141 and 142 and between weeks 206 and 207, which have been labeled as the week 142 and week 207 breakpoints, respectively. The betweenness centralization model yielded only one breakpoint at week 54, while entropy gave eight breakpoints at weeks 7, 10, 42, 92, 93, 204, 250, and 335.

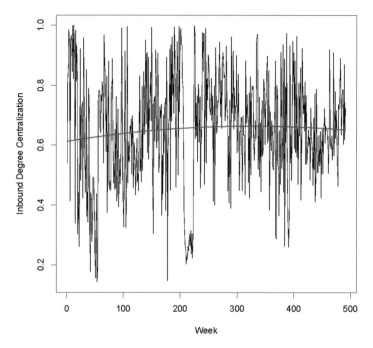

Fig. A.5 Plot of inbound degree centralization raw data and its fitted values from the regression model (Table A.4)

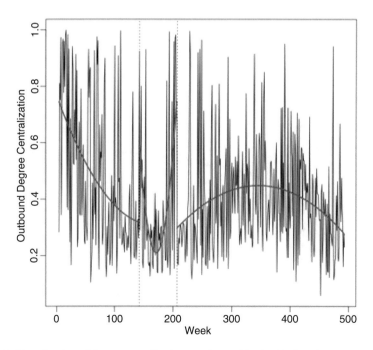

Fig. A.6 Plot of outbound degree centralization raw data and its fitted values from the regression model (Table A.5), with breakpoint locations marked with *dotted lines*

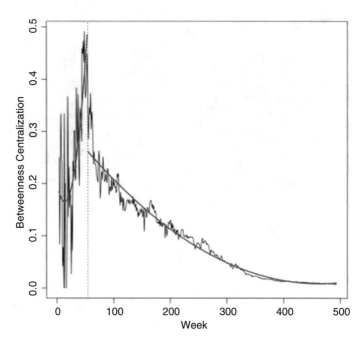

Fig. A.7 Plot of betweenness centralization raw data and its fitted values from the regression model (Table A.6), with the breakpoint location marked with a *dotted line*

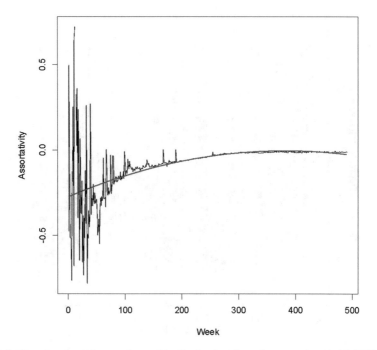

Fig. A.8 Plot of assortativity raw data and its fitted values from the regression model (Table A.7)

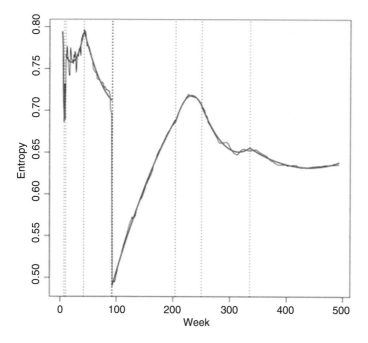

Fig. A.9 Plot of entropy raw data and its fitted values from the regression model (Table A.8), with breakpoint locations marked with *dotted lines*

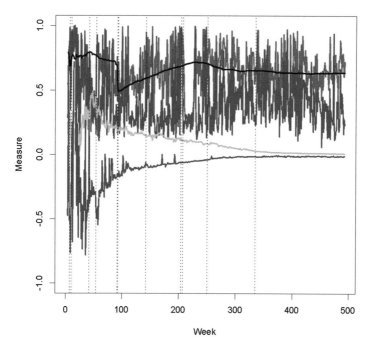

Fig. A.10 Plot of inbound degree centralization, outbound degree centralization, betweenness centralization, assortativity, and entropy raw data, with breakpoint locations marked with *dotted lines*

References

Britt BC (2011) System-level motivating factors for collaboration on Wikipedia: a longitudinal network analysis. Purdue University, Thesis

Britt BC (2013) Evolution and revolution of organizational configurations on Wikipedia: a longitudinal network analysis. Dissertation, Purdue University

Crawley MJ (2007) The R book. Wiley, West Sussex

de Alfaro L, Adler B, Pye I (2010) Computing wiki statistics (WikiTrust). http://wikitrust.soe.ucsc.edu/computing-wikistatistics. Accessed 10 Sept 2012

Freeman LC (1979) Centrality in networks: I. Conceptual clarification. Soc Networks 1:215–239

Isard W (1954) Location theory and trade theory: short-run analysis. Q J Econ 68:305–322

Lemoine N (2012) R for ecologists: putting together a piecewise regression. http://climateecology.wordpress.com/2012/08/19/r-for-ecologists-putting-together-a-piecewise-regression. Accessed 6 Jun 2013

Matei SA, Bertino E, Zhu M, Liu C, Si L, Britt BC (2015) A research agenda for the study of entropic social structural evolution, functional roles, adhocratic leadership styles, and credibility in online organizations and knowledge markets. In: Bertino E, Matei SA (eds) Roles, trust, and reputation in social media knowledge markets: theory and methods. Springer, New York, pp 3–33

Matei SA, Bruno RJ, Morris P (2010a) Visible effort: a social entropy methodology for managing computer-mediated collaborative learning. Paper presented at the Global Communication Forum, Shanghai, 29–30 Sept 2010

Matei SA, Oh K, Bruno R (2010b) Collaboration and communication in online environments: a social entropy approach. In: Oancea M (ed) Comunicare şi comportament organizational (Communication and organizational behavior). Printech, Bucharest, pp 82–98

Newman MEJ (2002) Assortative mixing in networks. Phys Rev Lett 89:208701

Purdue University (2008) Coates. http://www.rcac.purdue.edu/userinfo/resources/coates. Accessed 5 Mar 2013

Zeileis A, Kleiber C, Krämer W, Hornik K (2003) Testing and dating of structural changes in practice. Comput Stat Data An 44:109–123

Zeileis A, Shah A, Patnaik I (2010) Testing, monitoring, and dating structural changes in exchange rate regimes. Comput Stat Data An 54:1696–1706

Appendix B: Historical and Media Analysis of Wikipedia's Evolutionary Context

Introduction

An investigation of Wikipedia's evolution would not be complete without putting it in the context of real life events. While Wikipedians were adding, deleting, reverting, and debating, the world was moving apace, marked by some extremely important events. We recount them here for two reasons. First, in doing so, we provide some overall context for the historical era in which Wikipedia was born. Second, and more importantly, this effort connects some especially relevant developments to specific organizational changes, showing how societal and environmental forces may have served as evolutionary motors that influenced editors' behavior on Wikipedia, even if only indirectly, and demonstrating that they can be seen as feeding factors for editing increases and the expansion of Wikipedia's content overall.

Our historical review is therefore rather detailed. The goal is to provide, as has previously been mentioned, an idiographic description of the events at hand. Therefore, the scale of the described events is intentionally taken at the smallest level of granularity, as we aim to detail all of the idiosyncratic phenomena that surrounded Wikipedia's evolution.

We also offer this as tool for future research, providing a first layer for a future "thick description" (Geertz 1973) of the environment of events and ideas in which Wikipedia was born. Thus, what follows is a *sui generis* primary and secondary document, which may prove useful especially in future research. If the details tend to be many, the justification is that reality tends to be overabundant.

© Springer International Publishing AG 2017
S.A. Matei, B.C. Britt, *Structural Differentiation in Social Media*, Lecture Notes in Social Networks, DOI 10.1007/978-3-319-64425-7

Data Collection Strategy

We used the "Major Newspapers" archive within the LexisNexis Academic database as a source for prominent historical events. This particular resource offered two major benefits: the high-profile newspapers themselves and LexisNexis' own metadata about the articles in question.

First, in terms of the newspapers, the "Major Newspapers" grouping itself consisted of 69 of the most well-respected English-language media sources in the world. As described by LexisNexis (2013b), in order to be included within this archive:

> United States newspapers must be listed in the top 50 circulation in Editor & Publisher Year Book. Newspapers published outside the United States must be in English language and listed as a national newspaper in Benn's World Media Directory or one of the top 5% in circulation for the country.

All told, the range of this archive ensured that only significant topics that could realistically have an effect across Wikipedia coeditors would receive substantial attention, yet it also captured those issues that emerged both in the USA and worldwide.

Just as importantly, each article featured several pieces of key metadata that allowed the archive to be quickly reduced to only the most relevant potential forces. Each article was labeled based on its publication date, location in the newspaper, and key organizations, people, and subjects (or, to put it more generally, the article's key terms), as identified through LexisNexis' SmartIndexing Technology (LexisNexis 2013a). In effect, these key terms serve as tags, the results of a content analysis that LexisNexis conducted on its own articles.

These were used to conduct two separate data collection processes for newspaper articles for each breakpoint. For the first search, all front-page articles from the "Major Newspapers" archive within the 4 weeks prior to a given breakpoint and the week following it were retrieved from LexisNexis (see Chap. 10 for details on the time range). These records were further reduced to only the most prominent recurring terms using an R script that counted the number of times each subject appeared in all of the front-page articles for a given breakpoint period; created three scree plots of the number of appearances of each organization, person, and subject; and found the optimal cutoff to define highly significant terms for each of these plots. This cutoff point was determined by iterating through each term and, at each iteration, creating a regression model of two linear segments joined at the location of that term on the plot's x-axis, determining the error sum of squares for that regression model, and selecting the model that yielded the lowest error sum of squares.

For the selected model, the term corresponding to the location of the bend or "elbow" in the scree plot was deemed the ideal cutoff point for prominent terms of that category (organizations, people, and subjects), and all terms with at least as many front-page appearances as the identified cutoff term were treated as prominent. This approach was based upon the use of scree plots to identify key factors in principal component analysis, with the idea that there is a distinction between the

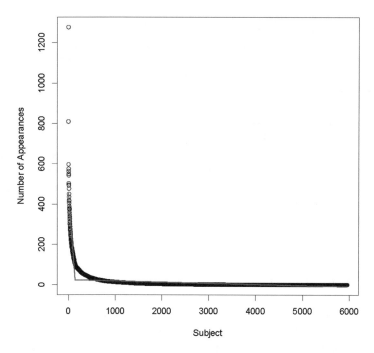

Fig. B.1 Scree plot for subjects featured in weeks 10–42, with the number of appearances of each subject and the corresponding segmented regression line

mass of news topics that receive brief, dispersed attention, and those that either receive sustained front-page coverage in a few markets or a burst of front-page coverage around the world.

Figure B.1 shows one of these scree plots, with subjects ordered from most appearances to least appearances. All subjects left of the "elbow" point in the regression line were considered prominent enough to warrant further examination.

Just as the 5-week time range was intentionally devised to be wider than was likely necessary, it was expected that many of the "prominent" terms within that range would not be remotely connected with the behavior of Wikipedia editors. With that in mind, this approach did not serve as a final analysis to define a concrete set of relevant terms; rather, it was instead used to filter approximately 100,000 front page "Major Newspapers" articles into a short list of those events and issues that had a major societal impact and which may therefore have impacted Wikipedia contributors as well, to be used in the next analysis step.

As for the second data collection process, in addition to the front-page articles that were retrieved regardless of subject, another search was conducted to retrieve all articles within a given breakpoint period, front-page or otherwise, from the "Major Newspapers" database, as long as they contained the word "Wikipedia" somewhere within the text. This produced a much smaller number of records than the thousands of front-page articles initially retrieved for each period. For the sake of consistency, the subjects addressed in each article mentioning Wikipedia were ordered by number of appearances in the same way as the front-page articles, but the smaller set of results was not further refined through the use of a scree plot. The topics uncovered through this second mode of data collection are addressed at greater length in Chap. 10.

We examined the full text of each article that matched a prominent term or subject word. For each breakpoint, we identified the major news items that coalesced into key themes, whether they dealt with momentous events like a presidential inauguration or a school shooting, or a very long-term news item like economic woes or the war on terror. For the sake of space, not all terms that were identified as prominent within the scree plot are discussed at length within this written document, but any terms that were found to be related to a given configurational change are indeed included in full.

First Breakpoint: Week 7

Wikipedia was born in 2001, a year of momentous change and dramatic shifts in American politics. The year started with a lingering controversy surrounding the presidential election and, following the 9/11 terrorist attacks, ended with a major military conflict in Afghanistan. The major events surrounding week 7 of Wikipedia's existence were of a largely political nature. It would be reasonable to speculate that these fueled some of the activity on the site, particularly since Wikipedia did provide some information about the election and the vote count.

It is worth mentioning that during the 10th week of Wikipedia's life, a controversial Supreme Court ruling halted a recount of Florida ballots from the 2000 US Presidential Election, so although some voters held lingering doubts about what the recount results would have been, George W. Bush took office as the 43rd President of the USA on January 20, 2001. Arguments over the election results continued well into Bush's first term as president. Several newspapers jointly conducted a partial, unofficial recount of some of the questionable ballots (Cauchon and Drinkard 2001), and a *New York Daily News* headline from February 26, 2001, ultimately proclaimed "Miami-Dade Recount BUSH REALLY WON New tally nets Gore only 49 votes" (Dineen 2001). Nevertheless, a pall remained over much of Bush's time in office, as some citizens argued that the Supreme Court effectively handed Bush the election and wrongly mitigated the public's role in the democratic process, even as most media outlets shifted their attentions to Bush's policies as the controversy subsided.

With that said, in the weeks that followed, a great deal of the media's focus was diverted toward the case of FBI agent Robert Hanssen. On February 18, 2001, Hanssen was arrested for selling American secrets to the Soviet Union and Russia (Savino 2001; Willing and Watson 2001). Hanssen reportedly made more than $1.4 million for his spying activities from 1979 to 2001, and he would ultimately plead guilty several months after his arrest (Risen 2001).

The embattled music-sharing service Napster stood alongside Hanssen in legal news. Several record labels along with the RIAA had already won a 2000 copyright infringement lawsuit against Napster, but its owners continued the fight in the US Court of Appeals. Napster was ultimately allowed to continue operating, but only if the system tracked user activity and if its owners restricted access to copyright-infringing material upon being notified of its presence. The company was only able to guarantee that it would be able to restrict 99.4% of copyright-infringing activity, which the court said was not good enough, so Napster was eventually closed in July 2001 (Douglas 2004; Lessig 2004).

Europe was facing its own share of problems, as a plague of foot-and-mouth disease swept through the continent (Cryderman 2001; Kennedy 2001). The first cases of foot-and-mouth were discovered in British cattle in mid-February, and the disease quickly spread into France and other parts of Europe. Other countries were compelled to enforce additional airport screenings for passengers arriving from Europe, and many placed temporary embargoes against European agricultural products in order to protect their own supplies. The disease was not contained until October 2001, by which time Great Britain had slaughtered over ten million sheep and cattle in its efforts to control the spread (Uhlig 2002). All told, the plague cost the UK an estimated £8.5 billion ($12.9 billion) in 2001 (Cormier 2013).

Overall, between January 30, 2001, and March 26, 2001, a total of 1,616 organizations, 4,934 people, and 8,439 subjects appeared as key terms within the "Major Newspapers" database. These lists were reduced to the 27 organizations, 46 people, and 192 subjects that appeared in the most articles based on the results of our three scree plots, and all articles featuring at least one of these especially prominent terms were analyzed to determine the most important recurring themes in the media.

Second Breakpoint: Week 42

This breakpoint, which appeared during the first week of November, came shortly after the 9/11 terrorist attacks. One might speculate that some of the work on Wikipedia could have been impacted by the "canary in the coalmine" effect wherein people demanded more answers and Wikipedia editors strived to offer them, ultimately attracting more contributors, more data, and more information. Case in point: Over the 4 weeks preceding 9/11, Wikipedia attracted 20 new authors, reaching a total of 145. In the 4 weeks after the attack, on the other hand, Wikipedia recruited 84 new authors, reaching a total of 229. Similarly, the number of articles revised jumped from just under 7,000 to over 11,000, and the number of revisions likewise increased by almost 50% (Appendix A, Table A.1). This is a significant outpouring of new activity.

Of the many other issues that dominated the public conversation during this week, a few topics accrued scattered attention in the press, like the 2001 Australian federal election, which saw Prime Minister John Howard's Liberal Party, alongside their coalition partners, the National Party of Australia, defeat the Australian Labor

Party headed by Kim Beazley (Australian Electoral Commission 2013; Lewis 2001; Macken 2001). And in baseball news, Barry Bonds broke the all-time single-season home run record on October 5, 2001, amidst a barrage of questions about possible steroid use (Aratani 2001).

Yet the biggest topics were, of course, those related to terrorism. Beginning on September 18, several lethal anthrax-laced envelopes were delivered to unsuspecting recipients, killing five people and eventually shutting down post offices across the country (Beach 2001). The nation remained on high alert for quite some time in anticipation of further attacks, especially considering the unexpected use of America's own planes and postal service to attack its citizens where they lived (Beach and McCabe 2001). The FBI would conclude in 2010 that a single biologist, Bruce E. Ivins, prepared and mailed the anthrax spores himself, without any assistance from others, although some skeptics in Congress maintained that they did not believe the real culprit had been found. Regardless, Ivins committed suicide in 2008 as the government was preparing to indict him, so the FBI closed the case with the 2010 report (Warrick 2010).

The week 42 breakpoint fell in a media period dominated, according to our analysis, by 2,646 organizations, 6,981 people, and 18,435 subjects. These were reduced to the most prominent 16 organizations, 29 people, and 254 subjects that appeared repeatedly across articles.

Third Breakpoint: Week 54

Wikipedia's one-year anniversary heralded a dramatic change of its own in the form of the week 54 breakpoint. Prior to the archival analysis, the set of key terms was again reduced to only the most prominent recurring topics, with the full set of 976 organizations, 2,780 people, and 5,962 subjects condensed to the most frequently appearing 14 organizations, 20 people, and 145 subjects.

The theme of interest was again terrorism along with the major economic upheaval produced by the Enron scandal. On December 22, 2001, 28-year-old Richard Reid attempted to board an airplane in France with the soles of his sneakers stuffed with triacetone triperoxide, an explosive powder that experts said may have doomed the Boeing 767 aircraft along with its 183 passengers and 14 crew members (Hagel et al. 2001; Reid 2001). Thanks to flight delays and rainy weather, however, Reid struggled to ignite the explosive while on the plane, and after it became clear that he was trying to detonate his shoes, several passengers jointly subdued him. Reid pled guilty on October 4, 2002, and was sentenced to life in prison without the possibility of parole (Butterfield 2002). His actions contributed to tighter security screening procedures at airports, many of which remain in effect a decade and a half later (Leiser and Wilson 2002).

The story of John Walker Lindh, the American who went to Afghanistan to join al Qaeda, also began to unfold at the start of 2002. In May 2001, Lindh went to Afghanistan and decided to fight for the Taliban, readily engaging in combat with

American troops and their allies in Afghanistan. On November 25, 2001, Lindh's unit surrendered, and he was captured along with his comrades. After a failed prison uprising at Qala-i-Jangi and roughly 45 days of interrogations, Lindh was brought back to the USA to stand trial (Beach 2002); he pled guilty several months later and received a 20-year prison sentence (Liptak 2007).

Also at the beginning of 2002, the Enron Corporation was imploding as the result of one of the largest accounting scandals of all time. For years, its executives had used an array of accounting loopholes and complicated financial reports to overreport earnings and hide billions of dollars in debt from the company's board of directors and stockholders (Oppel and Labaton 2002). As a result, Enron's stock swelled to $90 per share in 2000 before diving below $1 by November 2001. The company filed for bankruptcy in 2001 and failed entirely shortly thereafter. At the time, it was the largest company in history to fall into bankruptcy. Enron executives and several involved parties eventually faced criminal charges for fraud, conspiracy, money laundering, and other related counts; 16 of them pled guilty to their respective charges, and another five were convicted at trial (Pasha and Seid 2006).

Fourth Breakpoint: Week 92

For this breakpoint we identified 1,216 organizations, 3,584 people, and 5,971 subjects that appeared at least once, which we then reduced to the set of 6 organizations, 9 people, and 136 subjects with the most frequent appearances. In the news, a number of serious and, in a few cases, deadly stories took command of the front pages. The war on terror still remained the most prevalent topic in the media, but with a rather different focus than in the previous year. Specifically, by the one-year anniversary of the September 11 attacks, most of the media's attention had refocused on America's efforts to combat Iraqi dictator Saddam Hussein. In mid-September, President Bush, along with Britain Prime Minister Tony Blair and Australian Prime Minister John Howard, warned Americans that Hussein would have access to nuclear weapons by Christmas (Barker 2002; Maguire 2002; Taylor 2002), and they contended that since an Iraqi attack on the USA was inevitable, they needed to take down the "murderous tyrant" first (Mackenzie 2002). Hussein offered to readmit U.N. arms inspectors for the first time in four years, supposedly to prove that his forces had no nuclear capabilities. Bush, however, dismissed the gesture as "a cynical ploy" designed to stall the implementation of tougher U.N. resolutions (Harnden 2002). In addition to the nuclear threat, federal health officials were also preparing for the possibility of a bioterrorism attack that would reintroduce the smallpox virus to America (Auge and Sherry 2002; Franscell 2002).

Nonetheless, some analysts were beginning to blame the weakening US economy on Bush's "Iraq rhetoric" (Collins and James 2002), with reluctance toward the war (Knickerbocker 2002) and skepticism about Iraq's weapons of mass destruction (Black and von Sternberg 2002) beginning to spread through the country. The other members of the U.N. security council, especially France and Russia, expressed similar sentiments in response to a proposed resolution which would have authorized an invasion of Iraq, chastising US representatives for going "against international law" (Overington 2002).

During the same period, the USA was preparing for its midterm elections. Prior to the campaign, Democrats had hoped to strengthen their control over the Senate, particularly given the number of Republican-held seats that would be contested and the stumbling national economy. However, with the November 5 elections coming only 14 months after the September 11 attacks, Democrats struggled to muster much momentum against their Republican rivals, particularly with a Republican sitting as president. Many races were too close to call until election day (Branch-Brioso and Plambeck 2002). In the end, Democrats would lose two Senate seats to the Republican Party, conceding control of the Senate along with the already Republican-controlled House of Representatives (Nagourney 2002). On the other hand, in the gubernatorial elections, Democrats gained three governorships while the Republicans lost only one (Fox News 2002).

This period also circumscribed the Beltway snipers' 3-week reign of terror. John Allen Muhammad and Lee Boyd Malvo carried out a series of shootings from October 2 through October 22, killing ten people and critically injuring three others around Washington, D.C. In each shooting, the victim was killed by a single bullet fired from long range. After Muhammad and Malvo killed four people in less than 2 h, fear pervaded the region. Parents forbade their children from riding the school bus home; schools were locked down, with no outdoor gym classes or recesses; and pedestrians zigzagged while crossing streets just to counteract the mere possibility that a sniper might be aiming at them (MacLeod 2002; see also Washington Post Staff Writers 2002). Police eventually connected the pattern of shootings and traced an earlier incident to Malvo, who they quickly discovered held close ties to Muhammad. The pair were arrested, and while Malvo, who was 17 at the time of the attacks, was sentenced to life in prison, Muhammad was sentenced to death and eventually executed in 2009 after multiple appeals were denied (Chang et al. 2002; White and Glod 2009).

Elsewhere, Russia was dealing with a hostage crisis, as Chechen rebels took control of a packed theater in Moscow on October 23 and held 912 Russians and foreign citizens captive (Krechetnikov 2012; Strauss 2002). After a 3-day standoff, Russian forces seized control of the building and killed the rebel chief (Engleman 2002), but at the cost of also killing 129 of the hostages—not with guns, but with an unknown chemical agent (Coker and Yermolenko 2002; Johnston 2013). The particular gas that Russia used was never revealed to the public, nor to foreign embassies in Russia that sought to offer their support, all of which complicated doctors' efforts to treat the victims and roiled the international press (Hunter 2002; Ingram 2002).

Yet the biggest story in the worldwide media may have been the bombing attack that killed 202 people and injured 240 more in Bali on October 12 (BBC 2003). In a coordinated effort, a suicide bomber blew up a backpack filled with explosives inside Paddy's Pub. Seconds later, another suicide bomber illegally parked a minivan, which held 12 filing cabinets containing a volatile chemical compound, outside the Sari nightclub across the street from Paddy's Pub and detonated it. The second blast decimated three buildings and ripped apart the surrounding area, leaving a one meter-deep crater in the street (Callinan 2002b; Moor 2012; Parkinson 2002).

This incident, which was eventually connected to Islamist extremists with ties to Osama bin Laden, furthered developing fears that Indonesia had become a hiding place for al Qaeda and other terrorist cells (Callinan 2002a; Firdaus 2002). Three of the terrorists were executed by a firing squad in 2008, and notably, shortly before their execution, Indonesia's Communication and Information Minister Mohammad Nuh finally asked his country's media outlets to stop calling the three executed terrorists "heroes" (Allard and Murray 2008; Antara News 2008).

Fifth Breakpoint: Week 142

The media analysis retained a set of 1,255 organizations, 3,917 people, and 5,723 subjects, which were condensed to the most prominent 18 organizations, 11 people, and 166 subjects, and once again, all articles featuring at least one of those prominent terms were studied within the further archival analysis. The week 142 breakpoint included the second anniversary of the September 11 terrorist attacks (Coorey 2003), but the period was otherwise uneventful compared with the US elections, the Bali bombings, and the Beltway sniper attacks observed during the analysis period for the previous breakpoint. However, it was apparent that anti-US sentiments had grown considerably over the previous year. French and German representatives criticized the USA for seeking greater U.N. involvement in the campaign against Iraq, especially given concerns about how quickly governmental control would be returned to the Iraqi people (Hartcher 2003b; Richburg 2003). US Democrats lambasted Bush for his plans to spend $87 billion more to rebuild Iraq, while other nations balked at his request for an additional $55 billion from them (Grice and Usborne 2003; Gumbel 2003; Hartcher 2003a; Hutcheson and Douglas 2003). The lack of evidence that Iraq ever possessed any weapons of mass destruction, and America's continued search for them, further amplified the anti-US sentiment abroad and the anti-Bush attitude among a growing number of Americans (Hartcher 2003b; Hutcheson and Thomma 2003; Miller 2003).

The US economy, at least, appeared to be on the road to recovery. US census data, released at the end of September, revealed that poverty rates rose in 2002 (Clemetson 2003), but the jobs recovery in the USA, along with similarly positive economic trends worldwide, served to offset the census findings (Aylmer 2003; Mellish 2003; Pretty 2003). California, on the other hand, was dealing with the aftermath of an energy crisis from 2000 to 2001, during which resource mismanagement hit the state with skyrocketing electricity costs and multiple large-scale blackouts. In 2003, Governor Gray Davis was targeted by a recall campaign by citizens who argued that he did not properly respond to the crisis, and after multiple delays, the state's first-ever recall election was scheduled for October 7 (Puzzanghera 2003). Arnold Schwarzenegger would eventually win the position of governor, claiming almost half the votes despite challenges from Democrat Cruz Bustamante, fellow Republican Tom McClintock, and 132 other candidates (Finnegan 2003).

Aside from the Iraqi campaign, the only real violence in the news stemmed from the long-standing conflict between Israel and its Palestinian neighbors. On September 9, two suicide bombings in Israel killed over a dozen people (Bennet and Myre 2003), and a second suicide bombing by a Syrian terrorist group reportedly claimed 21 lives on October 4. The next day, Israeli jets responded by bombing a terrorist base in Syria to send a "very clear message" to the terrorist organizations focusing on Israel as well as any countries that harbored terror cells (Blair and Stack 2003). They were met with limited support from the international community. While the EU formally declared Hamas, one of the key groups involved, a terrorist organization (Keinon 2003), the U.N. tried to enact a resolution that would have protected Palestinian leader Yasser Arafat from harm or expulsion by Israel—a resolution that the USA eventually vetoed in order to protect its Israeli allies (Teather 2003).

Congress also approved a do-not-call list to prevent unwanted telemarketing calls (McGough 2003), and there was growing controversy regarding the PATRIOT Act, a piece of legislation that offered special investigative privileges for the purpose tracking terrorist activity in the USA, a criterion that some commentators found overly broad (Lichtblau 2003). Otherwise, though, the period surrounding week 142 was relatively quiet compared with the news stories in proximity to previous breakpoints.

Sixth Breakpoint: Week 204

We retrieved 1,636 organizations, 4,971 people, and 5,672 subjects that appeared at least once during the period surrounding the week 204 breakpoint, and 33 organizations, 21 people, and 168 subjects were found to be especially prevalent. The biggest such story was undoubtedly the tsunamis that devastated Indonesia, Sri Lanka, India, and Thailand, along with numerous other countries in the region (Gardner and Schubert 2004). On December 26, a megathrust earthquake rocked the Indian Ocean off the coast of Sumatra; the undersea quake was the third largest ever captured on a seismograph, with its magnitude estimated to be above 9.0 (US Geological Survey 2011). Because of its location, the seismic activity itself did little damage. However, it triggered tsunamis that besieged a dozen different countries along the coast, annihilating entire communities in an instant. Taken together, the tsunamis that stemmed from the Sumatran quake constituted one of the deadliest natural disasters ever recorded, claiming the lives of an estimated 225,000–275,000 people (CBC 2010) and leaving millions more injured or homeless. The disaster prompted worldwide humanitarian aid efforts that ultimately exceeded $14 billion (Jayasuriya and McCawley 2010).

Iraq was facing problems of a different sort. Its first democratic elections were close on the horizon, yet clashes with insurgents in Fallujah (Allam and Landay 2004) and elsewhere in the troubled nation (Vick and Sebti 2004) threatened to derail the election entirely (Strobel et al. 2004). The Elections for the National Assembly of Iraq did indeed transpire on January 30, 2005, although there were more than 100 armed attacks on polling places. Still, tight security prevented any major disruptions despite extremist threats to "wash the streets in blood" (Associated Press 2005).

Finally, in the USA, the press was ardently covering the fallout of the Guantanamo Bay detention camp scandal. In June 2004, a Red Cross team that included experienced medical personnel visited the detention center, and late in November 2004, the team's report, which alleged significant detainee abuses (Leonnig 2004), was released (Lewis 2004).

Seventh Breakpoint: Week 250

Our search in the proximity of week 250 retrieved 1,101 organizations, 3,049 people, and 4,628 subjects, lists that were subsequently reduced to the most frequently recurring 20 organizations, 19 people, and 167 subjects.

Sadly, this breakpoint coincided with another large-scale natural disaster. On October 8, a major earthquake struck Kashmir, a disputed territory administered by India, Pakistan, and China. The earthquake's strength was recorded as magnitude 7.6 on the Richter scale, and an estimated 100,000 people were killed as a result of the quake. 138,000 more were seriously injured and approximately 3.5 million lost their homes (Haseeb et al. 2011; Mirror Australian Telegraph Publications 2005a, b). The earthquake was so severe that it even cracked some of the nearby mountains (Sengupta and Rohde 2005).

An estimated \$6.7 billion in international aid was pledged toward the relief efforts (Associated Press 2010), but much like relief efforts for the Sumatran tsunamis a year earlier, help was slow to arrive to survivors. In this case, however, the delays were blamed more on political strife and governmental mismanagement of resources than on the conditions themselves. Many of the best-organized aid groups were radical Islamic factions, some of whom had thrown their support behind the Taliban. Pakistan's forces, on the other hand, devoted much of their energy toward fending off the Taliban along the Afghanistan border, so the groups with the best opportunity to assist in the relief effort on the Pakistani side of the border were reportedly fighting one another instead. Likewise, India was reportedly slow to assist its own citizens in the wake of the disaster, leading some to question the benefit of being Indian (Baldauf and Winter 2005).

In Iraq, the big story was the beginning of Saddam Hussein's criminal trial (Ford 2005; Mirror Australian Telegraph Publications 2005c). The defiant former dictator was charged with murder, genocide, and ethnic cleansing before the Iraqi Special Tribunal on June 30, 2004, based on his actions in the 1982 Dujail massacre (Spinner 2005). Hussein, who had to be held in a bulletproof cage during the trial, asserted that he was still the President of Iraq and that the court had no authority to judge him. Arguments about the court's validity notwithstanding, Hussein would be convicted of crimes against humanity on November 5, 2006 and hanged on December 30, 2006 (BBC 2006).

The American press, for its part, was just as concerned with the aftermath of Hurricane Katrina, which ravaged the country's eastern coast in late August. Some parties were working to assist with the recovery efforts, whether by providing housing (Gurnon 2005) or offering employment (Armour 2005). Others were more focused on determining who deserved blame for delays and resource shortages in the relief efforts (Grier 2005). Numerous aspects of the relief efforts were questionably managed, including $5.3 million in food donations that could not be delivered due to US rules (Connolly 2005), contracts that were awarded to large corporations instead of local businesses without any bidding process (Weisman and Witte 2005), and $200 million in donations to the Red Cross that were given for Katrina relief and that the organization instead earmarked for use in future crises (Crary 2005). The Federal Emergency Management Agency (FEMA) drew especially great public scorn for the relief efforts, which eventually led to the ouster of FEMA Director Michael Brown (Hsu 2005; McCaffrey 2005).

Eighth Breakpoint: Week 335

Week 335, the moment defining our final breakpoint, fell in mid-June 2007. A total of 463 organizations, 593 people, and 3,634 subjects appeared as key terms at least once in the "Major Newspapers" database for the 5-week analysis period, but only 18 organizations, eight people, and 124 subjects were found to hold special importance.

The Palestinian struggle between Hamas and Fatah was making headlines, with the Gaza Strip in the midst of a civil war between the two sects (Prusher 2007a, b). The conflict itself was nothing new; in 2006, Hamas had won control over the Palestinian Legislative Council, sparking political and military battles between the two groups. These clashes came to a head at this time, as the Battle of Gaza erupted from June 7 to June 15 (BBC 2007). As Hamas wrested command over the region from Fatah (Nissenbaum 2007), Palestinian President Mahmoud Abbas, a member of the Fatah party, dissolved the Palestinian government on June 14, declaring a state of emergency and dismissing Hamas-supported Prime Minister Ismail Haniyeh (Fisher and Wilson 2007; Wilson 2007). The two groups effectively split into two separate, autonomous governments, complicating international relations. Tensions in the region only grew worse when Hamas launched a fresh volley of rocket attacks at Israel, threatening to destabilize the entire region (Katz 2007; Katz and Keinon 2007; Katz et al. 2007).

During the same period, another global health scare stirred up public fears once more. An Atlanta man who had just returned from his honeymoon in Europe was quarantined on May 31 with a potentially deadly strain of tuberculosis (Young 2007). Many were appalled that Andrew Speaker was able to reenter the USA on May 24 despite his symptoms and the fact that European physicians had already directed him not to travel. The fear was exacerbated by the fact that computer screening at the border failed to properly result in Speaker's quarantine, stoking concerns that the imperfect system might be allowing other dangerous illnesses into America's borders as well (Bluestein and Barrett 2007; Deans 2007; Neergaard and Barrett 2007).

Finally, the Group of Eight (G8) summit convened from June 6 to 8 to discuss the question of climate change. While the G8's European members hoped to set specific, binding benchmarks for greenhouse emission reduction, US representatives balked at the idea of aligning its standards solely with those of the other European members, saying that China and India had to be part of any global climate change plan (Adam and Wintour 2007; Blair and Helm 2007). China's proposal, in contrast, explicitly prioritized economic development and poverty reduction over controlling greenhouse gases, eschewing any quantitative requirements for emission reduction (Jiangtao 2007). The summit eventually resulted in a climate deal that offered no firm targets, only a promise to "seriously consider" setting a goal of halving emissions by 2050 (Laghi 2007).

It was little wonder that the parties failed to reach a compromise on climate change, especially given arguments over Russia's proposed missile defense shield. Russian President Vladimir Putin had spent months rejecting a White House plan to build a missile defense base in Eastern Europe. He assailed the USA for its plans to build missile defense bases in the area he still considered a Russian zone of influence. British Prime Minister Tony Blair, in turn, warned Putin that European businesses would leave Russia unless the country embraced western values, to which Putin replied that he was the world's "only true democrat" (MacLeod 2007; Saunders 2007).

Concluding Remarks

This overview of world events highlights the variety of sociohistorical and intellectual influences that defined the context in which Wikipedia grew. Although the results articulated in Chap. 10 indicate that most of the transitions observed were likely triggered primarily by Wikipedia-specific internal factors and external influences, our detailed investigation on the events surrounding these breakpoints illuminates the fact that Wikipedia's growth did not occur in a historical void. Rather, the community was surrounded by major historical events, especially major natural disasters and momentous civil and military conflicts. While we cannot propose any direct causal relationships between these events and Wikipedia collaborative patterns, we would like to highlight the fact that the growth of the project, especially its agglutination of information in a short period of time, can and does account for these events. At each turning point, major events created nodes of interest and activity that, at the very least, fed the public's hunger for information and fostered subsequent activity on Wikipedia. The precise manner in which these events interacted with Wikipedia's development, however, demands further investigation.

References

Adam D, Wintour P (2007) Climate change: new global plan to tie in worst polluters: Britain and Germany led effort to get US, China and India to agree to carbon trading scheme. The Guardian, p. 1

Allam H, Landay J (2004) Conflict in Iraq Fallujah to have nasty effects; Government will pay political price ahead of January elections. Pioneer Press, Saint Paul, p A1

Allard T, Murray L (2008) Torrent of rage as Indonesia on high alert. http://www.smh.com.au/news/world/torrent-ofrage-as-indonesia-on-high-alert/2008/11/09/1226165386706.html. Accessed 3 July 2013

Antara News (2008) Minister asks media not to call Amrozi et al: Heroes. http://www.embassyofindonesia.org/news/2008/10/news152.htm. Accessed 3 July 2013

Aratani L (2001) Fan launches legal skirmish for record Bonds ball: Berkeley man claims he caught it before scuffle. Mercury News, San Jose, p 1A

Armour S (2005) Evacuees finding jobs away from Katrina territory. USA Today, p 1A

Associated Press (2005) Iraq polling stations under attack. The Guardian. http://www.guardian.co.uk/world/2005/jan/29/iraq1. Accessed 3 July 2013

Associated Press (2010) Study finds that foreign aid after 2005 Kashmir earthquake built trust in Pakistan. Fox News. http://www.foxnews.com/world/2010/09/07/study-finds-foreign-aid-kashmir-earthquake-built-trust-pakistan. Accessed 3 July 2013

Auge K, Sherry A (2002) Smallpox vaccine plan would use polling places; Colo. reacts to U.S. call for mass shots within week of any outbreak. The Denver Post, p A-01

Australian Electoral Commission (2013) 2001 Federal Election. http://www.aec.gov.au/elections/federal_elections/2001/index.htm. Accessed 1 July 2013

Aylmer S (2003) US economy buoyed by jobs recovery. The Australian Financial Review, p 1

Baldauf S, Winter LJ (2005) Kashmir prized but little aided. Christian Science Monitor, p 01

Barker G (2002) Howard reveals Iraq's weapons plans. The Australian Financial Review, p 1

BBC (2003) Bali death toll set at 202. BBC. http://news.bbc.co.uk/2/hi/asia-pacific/2778923.stm. Accessed 2 July 2013

BBC (2006) Saddam Hussein executed in Iraq. BBC. http://news.bbc.co.uk/2/hi/middle_east/6218485.stm. Accessed 3 July 2013

BBC (2007) How Hamas took over the Gaza strip. BBC. http://news.bbc.co.uk/2/hi/middle_east/6748621.stm. Accessed 3 July 2013

Beach M (2001) Anthrax invades America – Mailrooms close after fourth case confirmed – War against terror. The Sunday Telegraph, p 1

Beach M (2002) America's traitor faces the music – Shaven and cuffed Walker for trial in US. The Daily Telegraph, p 1

Beach M, McCabe H (2001) US warned more terrorist attacks on way. The Daily Telegraph, p 1

Bennet J, Myre G (2003) Double suicide bombings rock Israel: Sharon cuts trip short as 13 killed. The Gazette, p A1

Black E, von Sternberg B (2002) Confronting Iraq: as President Bush builds his case against Iraq, America is grappling with the implications. Star Tribune, p 1A

Blair D, Helm T (2007) US rejects Blair's deal on carbon dioxide emissions. The Daily Telegraph, p 1

Blair D, Stack MK (2003) Israeli jets bomb terror base in Syria: expansion of war 'sends message' to enemies. The Ottawa Citizen, p A1

Bluestein G, Barrett D (2007) TB patient warning ignored at border; U.S. inspector says infected man appeared perfectly healthy. Pittsburgh Post-Gazette, p A-1

Branch-Brioso K, Plambeck J (2002) Contests are "a nail-biter" across U.S. St. Louis Post-Dispatch, p A1

Butterfield F (2002) Qaeda man pleads guilty to flying with shoe bomb. The New York Times. http://www.nytimes.com/2002/10/05/national/05SHOE.html. Accessed 1 July 2013

Callinan R (2002a) Bali bombers paid by bin Laden. Courier Mail, p 1

Callinan, R (2002b) Bus bomb – Revealed: How the Bali attack was staged – Terror in Bali. The Daily Telegraph, p 1

Cauchon D, Drinkard J (2001). Florida voter errors cost Gore the election. USA Today. http://usatoday30.usatoday.com/news/politics/2001-05-10-recountmain.htm. Accessed 1 July 2013

CBC (2010) The world's worst natural disasters. CBC. http://www.cbc.ca/news/world/story/2008/05/08/f-naturaldisasters-history.html. Accessed 3 July 2013

Chang D, McCaffrey S, Merzer M (2002) Police: sniper case is solved: tests link rifle found in suspects' car to 11 of 13 shootings, officials say Gulf War veteran, teen are surrounded asleep at rest stop. St. Louis Post-Dispatch, p A1

Clemetson L (2003) More Americans in poverty in 2002, census study says. The New York Times, p A1

Coker M, Yermolenko I (2002) Gas used in Russian rescue raid, over 140 die. The Atlanta Journal-Constitution, p 1A

Collins L, James C (2002) US markets slump as Bush steps up Iraq rhetoric. The Australian Financial Review, p 1

Connolly C (2005) Katrina food aid blocked by U.S. rules; Meals from Britain sit in warehouse. The Washington Post, p A01

Coorey P (2003) Remember – as the world pauses to reflect, evil reappears – 9/11 two years on – special edition. The Daily Telegraph, p 1

Cormier Z (2013) Synthetic vaccine could prevent future outbreaks of foot-and-mouth disease. Nature. http://www.nature.com/news/synthetic-vaccine-could-prevent-future-outbreaks-of-foot-and-mouth-disease-1.12700. Accessed 1 July 2013

Crary D (2005) Red cross under fire for Katrina efforts. Saint Paul Pioneer Press, p 1A

Cryderman K (2001) Simulated outbreak backs use of vaccine: in a foot-and-mouth 'war game,' slaughter was not enough. The Ottawa Citizen, p A1

Deans B (2007) Lawmakers rip CDC 'meltdown' in TB case. The Atlanta Journal-Constitution, p 1A

Dineen JK (2001) Miami-Dade recount Bush really won new tally nets Gore only 49 votes. New York Daily News, p 1

Douglas G (2004) Copyright and peer-to-peer music file sharing: the Napster case and the argument against legislative reform. http://www.murdoch.edu.au/elaw/issues/v11n1/douglas111.html. Accessed 1 July 2013

Engleman E (2002) Russian forces seize theater, kill Chechen rebel chief; 20 dead. Pittsburgh Post-Gazette, p A-1

Finnegan M (2003) Gov. Davis is recalled; Schwarzenegger wins. Los Angeles Times. http://articles.latimes.com/2003/oct/08/local/me-recall8. Accessed 3 July 2013

Firdaus I (2002) Bomb blast kills over 170 at resort nightclub in Bali; a second bomb explodes near U.S. consular office. St. Louis Post-Dispatch, p A1

Fisher M, Wilson S (2007) Defying Hamas, Abbas swears in new cabinet; move opens up new paths for peace, Israeli leader says. The Ottawa Citizen, p A1

Ford P (2005) Can Saddam Hussein get a fair trial? Christian Science Monitor, p 01

Fox News (2002) Governor races results. Fox News. http://www.foxnews.com/story/2002/11/06/governor-racesresults. Accessed 2 July 2013

Franscell R (2002) Smallpox 'holocaust' Indiana forebears' fate recalled amid specter of bioterror. The Denver Post, p A-01

Gardner S, Schubert M (2004) Tsunami toll predicted to pass 60,000; Thousands more missing massive relief operation Eight Australians dead. The Age, p 1

Geertz C (1973) The interpretation of cultures: selected essays. Basic Books, New York

Grice A, Usborne D (2003) US isolated as Europe scorns plea for more troops in Iraq. The Independent, p 1

Grier P (2005) Where to find $200 billion to pay for Katrina. Christian Science Monitor, p 01

Gumbel A (2003) US woos Europe as costs force Iraq u-turn. The Independent, p 1

Gurnon E (2005) Twin Cities public housing possible for Katrina victims; But preference would lengthen wait for needy local families. Saint Paul Pioneer Press, p 1A

Hagel J, Borenstein S, Chatterjee S (2001) Explosive in shoes endangered plane; Experts would expect a hold in fuselage, perhaps a crash. Saint Paul Pioneer Press, p A1

Harnden T (2002) Saddam's 'cynical' offer is rejected. The Daily Telegraph, p 1

Hartcher P (2003a) Bush faces $134bn Iraq bill blow-out. The Australian Financial Review, p 1, 12

Hartcher P (2003b) Iraq dragging Bush down. The Australian Financial Review, p 1

Haseeb M, Xinhailu BA, Khan JZ, Ahmad I, Malik R (2011) Construction of earthquake resistant buildings and infrastructure implementing seismic design and building code in northern Pakistan 2005 earthquake affected area. Int J Bus Soc Sci 2(4):168–177

Hsu SS (2005) Brown defends FEMA's efforts; Former agency director spreads blame for failures in Katrina response. The Washington Post, p A01

Hunter B (2002) Nerve gas mystery: uproar over Moscow hostage siege. The New York Post, p 001

Hutcheson R, Douglas W (2003) Bush team admits little Iraq help likely. The Philadelphia Inquirer, p A01

Hutcheson R, Thomma S (2003) Republicans suddenly fretful over Bush's chances in 2004; His popularity sinks as public's anxieties over Iraq, economy rise. Saint Paul Pioneer Press, p A1

Ingram J (2002) Doctors still in dark on fatal gas identity. Pittsburgh Post-Gazette, p A-1

Jayasuriya S, McCawley P (2010) The Asian tsunami. Edward Elgar Publishing, Northampton

Jiangtao S (2007) Beijing unveils plan to tackle climate change; Nation joins global warming fight. South China Morning Post, p 1

Johnston WR (2013) Worst terrorist strikes – worldwide. http://www.johnstonsarchive.net/terrorism/wrjp255i.html. Accessed 3 July 2013

Katz Y (2007) Seven Palestinians killed as IAF steps up strikes. Aircraft hit 11 Gaza targets in a day * Rockets slam into 2 Sderot homes * Another Hamas minister nabbed. The Jerusalem Post, p 1

Katz Y, Keinon H (2007) Israel rejects idea of Gaza truce. 'They won't get a prize for stopping the fire that they escalated' senior official says. The Jerusalem Post, p 1

Katz Y, Toameh KA, Stoil RA (2007) Israel braces for rockets on Ashkelon Beersheba. Sneh to 'Post': Iran behind upsurge in violence. The Jerusalem Post, p 1

Keinon H (2003) EU declares Hamas a terrorist organization. The Jerusalem Post, p 1

Kennedy M (2001) Canada bans farm goods from Europe: British foot-and-mouth outbreak hits France, threatens to spread farther. The Ottawa Citizen, p A1

Knickerbocker B (2002) Americans back Iraq war – warily. Christian Science Monitor, p 1

Krechetnikov A (2012) Moscow theatre siege: questions remain unanswered. BBC. http://www.bbc.co.uk/news/worldeurope-20067384. Accessed 2 July 2013

Laghi B (2007) Climate deal struck – with no firm targets; After day of tough negotiation, leaders vow to 'seriously consider' goal of halving emissions by 2050. The Globe and Mail, p A1

Leiser K, Wilson C (2002) Nation's airports are ready to start screening all bags but new security rules won't eliminate threat, some experts warn. St. Louis Post-Dispatch, p A1

Leonnig CD (2004) Further detainee abuse alleged; Guantanamo prison cited in FBI memos. The Washington Post, p A01

Lessig L (2004) Free culture: the nature and future of creativity. Penguin, New York

Lewis NA (2004) Red cross finds detainee abuse in Guantánamo. The New York Times. http://www.nytimes.com/2004/11/30/politics/30gitmo.html. Accessed 3 July 2013

Lewis S (2001) Beazley's $3bn poll gamble. The Australian Financial Review, p 1

LexisNexis (2013a) Academic user guide. http://www.lexisnexis.com/documents/academic/academic_migration/LexisNexisAcademicUserGuide-1.pdf. Accessed 21 June 2013

LexisNexis (2013b) Major Newspapers. http://www.lexisnexis.com.ezproxy.lib.purdue.edu/lnacui2api/results/shared/sourceInfo.do?csi=8422. Accessed 18 June 2013

Lichtblau E (2003) Patriot Act used to fight crime with no link to terror. Pittsburgh Post-Gazette, p A-1

Liptak A (2007) John Walker Lindh's buyer's remorse. The New York Times. http://www.nytimes.com/2007/04/23/us/23bar.html. Accessed 1 July 2013

Macken D (2001) Women turn their backs on Howard. The Australian Financial Review, p 1

Mackenzie H (2002) Bush warns Iraq will attack U.S.: get the 'murderous tyrant' first, president tells nation. The Ottawa Citizen, p A1

MacLeod C (2007) Blair's G8 warning to Putin: business will pull out of Russia. The Herald, p 1

MacLeod I (2002) Sniper fears paralyse Washington: pedestrians zigzag as police hunt for mysterious white van. The Ottawa Citizen, p A1

Maguire T (2002) Bomb by Christmas – Chilling warning on Iraq's nuclear timetable. The Daily Telegraph, p 1

McCaffrey S (2005) Ex-FEMA boss puts blame on Louisiana; Brown defends his response to Katrina, draws wide rebuke. Saint Paul Pioneer Press, p 1A

McGough M (2003) Do-not-call list put back on track fast; Its ultimate fate, however, rests with Supreme Court. Pittsburgh Post-Gazette, p A-1

Mellish M (2003) Budget boost paves the way for tax cuts. The Australian Financial Review, p 1

Miller G (2003) Bush admits 'no evidence' links Saddam Hussein to 9/11 attacks. Saint Paul Pioneer Press, p A1

Mirror Australian Telegraph Publications (11 Oct 2005a) 40,000 dead: we've lost a generation of children – Quake disaster. The Daily Telegraph, p 1

Mirror Australian Telegraph Publications (10 Oct 2005b) Death toll soars by thousands in Kashmir quake. The Australian, p 1

Mirror Australian Telegraph Publications (20 Oct 2005c) Saddam trial begins. The Daily Telegraph, p 1

Moor K (2012) Victims recount the horrors of the Bali bombing in Keith Moor's 10th anniversary special report. Herald Sun. http://www.news.com.au/national-news/victoria/victims-recount-the-horrors-of-the-bali-bombing/story-fndo4cq1-1226489407574. Accessed 3 July 2013

Nagourney A (2002) The 2002 elections: the overview: G.O.P. retakes control of the Senate in a show of presidential influence; Pataki, Jeb Bush, and Lautenberg win. The New York Times. http://www.nytimes.com/2002/11/06/us/2002-elections-overview-gop-retakes-control-senate-show-presidential-influence.html. Accessed 2 July 2013

Neergaard L, Barrett D (2007) How did he get back into the country? He had a serious form of TB and was told not to re-enter the U.S. His name was put on a watch list and given to U.S. border guards. Gap in system shows how vulnerable nation is to epidemic, experts say. St. Louis Post-Dispatch, p A1

Nissenbaum D (2007) Hamas seizes control from Fatah; Palestinians faced with confusion of dueling leaders. Pittsburgh Post-Gazette, p A-1

Oppel RA Jr, Labaton S (2002) Congress shreds Enron auditors. Pittsburgh Post-Gazette, p A-1

Overington C (2002) France, Russia castigate US invasion plan. Sydney Morning Herald, p 1

Parkinson T (2002) 'Bin Laden' voices new threat to Australia. The Age. http://www.theage.com.au/articles/2002/11/13/1037080786315.html. Accessed 2 July 2013

Pasha S, Seid J (2006) Lay and Skilling's day of reckoning. CNN. http://money.cnn.com/2006/05/25/news/newsmakers/enron_verdict/index.htm. Accessed 1 July 2013

Pretty M (2003) Many happy returns as global recovery lifts markets. The Australian Financial Review, p 1

Prusher IR (2007a) As Gaza unravels, Palestinians flee. Christian Science Monitor, p 1

Prusher IR (2007b) Palestinian split rattles region. Christian Science Monitor, p 1

Puzzanghera J (2003) Recall energizes Demos nationally; Anger can be potent electoral force. San Jose Mercury News, p 1A

Reid TR (2001) Sneaker bomb suspect: from thievery to Jihad; Mosque leader doubts he acted alone. The Washington Post, p A01

Richburg KB (2003) Allies cool to new Iraq plan; Chirac, Schroeder criticize proposal for bigger U.N. role. Pittsburgh Post-Gazette, p A-1

Risen J (2001) Ex-agent pleads guilty in spy case. The New York Times, p A-1

Saunders D (2007) Cold warrior Putin threatens to target Europe; Over sea bass and caviar, he vows to turn missiles westward, lashes out at NATO and insists he's world's only true democrat. The Globe and Mail, p A1

Savino L (2001) FBI says top agent spied for Russia; Robert P. Hanssen stands accused of passing along thousands of secret U.S. documents since 1985. The Philadelphia Inquirer, p A01

Sengupta S, Rohde D (2005) Quake widens rift in families across Kashmir. The New York Times, p A1

Spinner J (2005) Hussein faces tribunal today in first trial for actions in Iraq. The Washington Post, p A01

Strauss J (2002) Chechens take 700 hostage in Moscow theatre. The Daily Telegraph, p 1

Strobel WP, Walcott J, Landay JS (2004) Officials privately say Iraq election in peril. Saint Paul Pioneer Press, p A1

Taylor L (2002) Blair: Iraq's 45 minutes to zero. The Australian Financial Review, p 1

Teather D (2003) US vetoes UN call to protect Arafat. The Guardian, p 1

U.S. Geological Survey (2011) FAQ – everything else you want to know about this earthquake & tsunami. http://earthquake.usgs.gov/earthquakes/eqinthenews/2004/us2004slav/faq.php. Accessed 3 July 2013

Uhlig R (2002) 10 million animals were slaughtered in foot and mouth cull. The Telegraph. http://www.telegraph.co.uk/news/uknews/1382356/10-million-animals-were-slaughtered-in-foot-and-mouth-cull.html. Accessed 1 July 2013

Vick K, Sebti B (2004) Fighting spreads in Iraq: Rebels appear to open new fronts amid U.S. assault on Fallujah. Saint Paul Pioneer Press, p A1

Warrick J (2010) FBI investigation of 2001 anthrax attacks concluded; U.S. releases details. The Washington Post. http://www.washingtonpost.com/wp-dyn/content/article/2010/02/19/AR2010021902369.html. Accessed 3 July 2013

Washington Post Staff Writers (2002) For parents and students, safety first; Schools lock their doors, and some keep information scarce in fighting fear. The Washington Post, p A13

Weisman J, Witte G (2005) Katrina contracts will be reopened; No-bid deals questions on hill. The Washington Post, p A01

White J, Glod M (2009) Muhammad is executed for sniper killing. Washington Post. http://www.washingtonpost.com/wp-dyn/content/article/2009/11/10/AR2009111001396.html. Accessed 2 July 2013

Willing R, Watson T (2001) FBI portrays Robert Hanssen's double life: a 15-year paradox. USA Today, p 1A

Wilson S (2007) Abbas dissolves government as Hamas takes control of Gaza. The Washington Post, p A01

Young A (2007) Atlantan quarantined with deadly TB strain; CDC issues rare isolation order; air passengers warned. The Atlanta Journal-Constitution, p 1A

Appendix C: Advantages and Disadvantages of Stepwise Segmented Regression Analysis

In the context of discussing our breakpoint detection strategy, it is worth noting a few key strengths and weaknesses of stepwise segmented regression analysis compared to other methods. First and foremost, unlike other algorithms, stepwise segmented regression is able to account for continuous and discontinuous breakpoints alike and can detect them without the need for *a priori* specification. It is also more robust than competing approaches due to its basis in regression, which allows it to handle high-variance data as well as, to a degree, heteroscedastic variance.

Additionally, because this method is based in regression, stepwise segmented regression is further extensible to multidimensional time series. In our study, we assessed five factors over time: inbound degree centralization, outbound degree centralization, betweenness centralization, assortativity, and entropy. Here, five separate stepwise segmented regression analyses were conducted to assess these five factors. However, a single multivariate regression could have been undertaken to achieve the same goal. In our case, the present study stood as a validation of the method itself, among other objectives, so conducting five separate analyses and observing the differences in the results offered a superior means to observe the analysis in action across a range of variables. Other researchers using this method may prefer a multivariate regression instead, as that approach will allow them to minimize the chance of erroneously detecting multiple breakpoints in close proximity when only one revolution actually transpired since all available data will be incorporated into the sole analysis being conducted rather than being partitioned across multiple regression analyses.

On the other hand, while the ability to include regression terms of multiple orders (intercept, linear, quadratic, etc.) offers stepwise segmented regression considerable versatility, doing so may make it difficult to detect breakpoints that occur in close proximity to one another due to the lack of a unique solution. For instance, if four regression terms are used in an analysis but a pair of revolutions occurs only 3 weeks apart, then the model for that 3-week segment and the associated revolutions will inevitably be overfit.

© Springer International Publishing AG 2017
S.A. Matei, B.C. Britt, *Structural Differentiation in Social Media*, Lecture Notes in Social Networks, DOI 10.1007/978-3-319-64425-7

More generally, terms of different orders can easily become confounded with one another, particularly for relatively short line segments between breakpoints. For example, if even one data point in a short line segment is an outlier, then while the revolutions circumscribing that evolutionary period may be correctly identified, that period itself may not be properly fitted. As an example, if a given measure experiences a momentary dip that is shortly followed by a return to a high level, the short period during which the measure adopted a low level might be inappropriately modeled as a positive quadratic term competing with a negative linear term, when in reality the period was one of relative stability. One could even argue, in fact, that the breakpoints corresponding to weeks 142 and 207, as modeled for outbound degree centralization (Fig. C.1), circumscribed a period that could have been modeled in a relatively flat, almost linear fashion rather than as a sharp curvilinear trend.

Such problems are unlikely if, for instance, only an intercept and a linear term are included in the stepwise segmented regression model, but they can quickly become hazardous with the introduction of additional variables. Therefore, care must be taken in incorporating quadratic, cubic, and other higher-order terms into any stepwise segmented regression analysis, as the potential consequences of including unnecessary terms are greater than in traditional regression.

As a final note, the algorithm upon which stepwise segmented regression is conducted is computationally complex. At each iteration of the process, the algorithm must conduct a complete segmented regression analysis corresponding to every

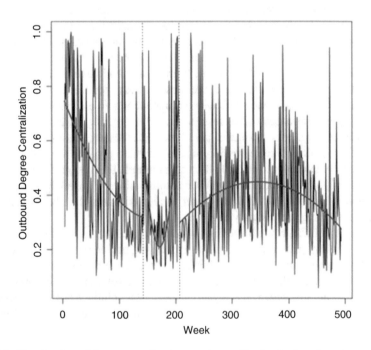

Fig. C.1 Plot of outbound degree centralization raw data and its fitted values from the regression model, with breakpoint locations marked with *dotted lines*

possible breakpoint that could be added to the model. As long as the data set contains only a few hundred or thousand data points, this is not a problem. However, a time series with millions of data points would be unwieldy, as this would effectively mean that millions of regression terms would have to be evaluated in the stepwise model selection process, which in turn would require millions of segmented regression analyses at every step. Again, this is not a problem for the vast majority of time series, for which stepwise segmented regression analyses can be conducted in minutes on an average PC. However, attempting to analyze an especially long time series with this approach would require either a degree of data compression or substantial modifications to the analysis itself, as even high-performance computing resources might struggle to handle the challenge of analyzing millions upon millions of successive segmented regression models.

Regardless of the limitations noted above, however, stepwise segmented regression analysis remains an optimal approach for analyzing longitudinal behavioral data, particularly due to its ability to identify any number of continuous and discontinuous breakpoints even amidst substantial error variance. As illustrated in Chap. 9, it is ideal for assessing large-scale community dynamics as they change over time, and it offers especially great potential for researchers exploring the array of online interaction data sets that are waiting to be mined.

Printed in the United States
By Bookmasters